EARTH and BEYOND

The latest view of the observed universe

ALEMU GONSAMO

Copyright © 2017 Alemu Gonsamo
All rights reserved.
ISBN-10: 1978342942
ISBN-13: 978-1978342941

*"Space: the final frontier. …
to explore strange new worlds,
to seek out new life and new civilizations,
to boldly go where no one has gone before"
Commanding Officer Jean-Luc Picard, Captain of the USS Enterprise*

TABLE OF CONTENTS

1. Earth

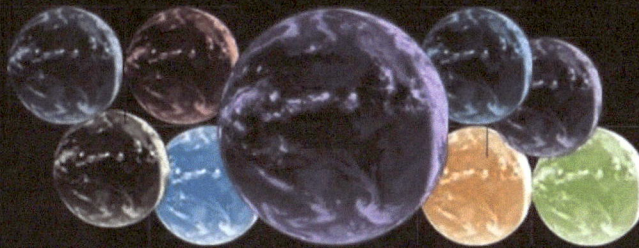
2. Our planets

3. Planetary system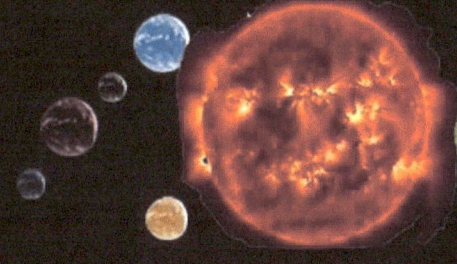

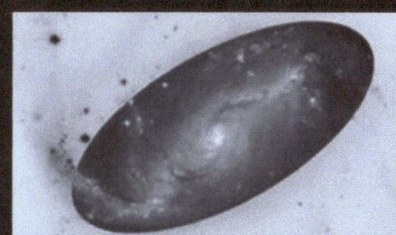
4. Galaxy

5. Universe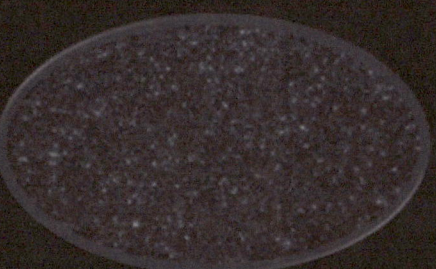

ACKNOWLEDGMENTS

The following sources are acknowledged for the materials presented in this book: National Aeronautics and Space Administration (NASA), NASA Goddard Space Flight Center, NASA Johnson Space Center, NASA Jet Propulsion Laboratory (JPL), NASA Imagine the Universe, European Space Agency (ESA), United States Geological Survey (USGS), The Hubble Heritage Team, Hubble Ultra Deep Field 2009 (HUDF09) Team, Deutsches Zentrum für Luft- und Raumfahrt e.V. (DLR), Association of Universities for Research in Astronomy (AURA), European Southern Observatory (ESO), Space Telescope Science Institute (STScI), Malin Space Science Systems (MSSS), Chinese Academy of Sciences, China National Space Administration (CNSA), CNSA Science and Application Center for Moon and Deepspace Exploration, Search for Extraterrestrial Intelligence (SETI) Institute, International Dark-Sky Association (IDA), Space Science Institute (SSI), Solar Dynamics Observatory, Chandra X-Ray Observatory (CXC), Wilkinson Microwave Anisotropy Probe (WMAP) Team, Rosetta NAVCAM Team, Philae CIVA Team, Microlensing Planet Search Project (MPS), Planck Collaboration, High Frequency Instrument (HFI) and Low Frequency Instrument (LFI) Consortia, GeoSphere Project, Caltech, Carnegie Institution of Washington, Southwest Research Institute (SwRI), Johns Hopkins University Applied Physics Laboratory, Cornell University, University of Arizona, University of Idaho, University of California-Los Angeles, Rochester Institute of Technology, Arizona State University, Tom Van Sant, Eugene Cernan, Kipp Teague, Glen Swanson, Eric M. Jones, Ken Glover, Emily Lakdawalla, J. Bell, M. Wolff, J. Nichols, A. Simon, M. Wong, G. Orton, Erich Karkoschka, S. Brunier, R. Lucas, Jeff Hester, Paul Scowen, J.Kastner et al., Andrew Z. Colvin, G. Illingworth, D. Magee, P. Oesch, R. Bouwens, J.Hester et al., Justin Cowart

Dr. Petra D'Odorico (PhD in Earth Observation), Dr. Sandro D'Odorico (PhD in Astrophysics) and Mr. Andrew Richard Malcolm (author and digital communications specialist) are acknowledged for their editorial contributions

1. Earth
1.1 A brief background of Earth

Earth is our planet, within our planetary system called Solar System which has a star called Sun, and located in our galaxy called Milky Way

Earth is the only known planet in our planetary system that is capable of sustaining life

Earth has one natural satellite called Moon. Moon is the only celestial body beyond Earth that has been visited by human beings

A natural satellite or Moon is a celestial body that orbits another celestial body of greater mass (e.g. a planet or star). Moon is a natural satellite of Earth, and Earth is a natural satellite of the Sun

Earth is 4.543 billion years old

Fact Box 1: launching and placing a spacecraft on an orbit

The world's first artificial satellite, the Sputnik 1, was launched by the Soviet Union in 1957. Since then, thousands of satellites have been launched into orbit around the Earth or other celestial bodies. Some satellites, notably space stations, have been launched in parts and assembled in orbit. Thousands of satellites are currently operational, whereas thousands of unused satellites and satellite fragments orbit the Earth as space debris. A few space probes have been placed into orbit around other celestial bodies and become artificial satellites to the Moon, Mercury, Venus, Mars, Jupiter, Saturn, Vesta, Eros, Ceres, and the Sun. Human spacecrafts in orbit are also satellites and satellite orbits vary greatly, depending on the purpose of the satellite, and are classified in a number of ways such as low Earth orbit, polar orbit, and geostationary orbit

Satellite components include spacecraft bus and payload. The spacecraft bus contains the support vehicle and control subsystems including: tracking, telemetry, command and ranging; solar panels to provide all of the prime power; batteries to operate the satellite when it is hidden from the Sun by the Earth or Moon; reaction control system (propulsion) to be used to correct the orbit; thermal control to maintain the electronics and other components within a safe temperature range over the life of the satellite; and structure to hold everything together and protect the subsystems during launch and after deployment on orbit. Payload is a module attached to a spacecraft bus containing the main data gathering instrument with communications circuitry that operates independently of the main spacecraft bus but which shares the power supply and transponders. Examples of payload include scientific instruments (cameras, telescopes, or particle detectors, for example), cargo, or a human crew

Satellites are propelled by launch vehicle or carrier rocket to their orbits. Usually the launch vehicle itself is a rocket lifting off from a launch pad on land, at sea, or aboard a plane. A launch system includes the launch vehicle, the launch pad, and other infrastructure. Earth orbital launch vehicles typically have at least two stages. Each individual stage of a rocket is generally assembled by mating of rocket stage(s) with the spacecraft body into a single assembly known as a space vehicle. Typically, launch vehicles are designed for one-time use and they usually separate from the satellite and disintegrate during atmospheric re-entry. Isaac Newton stated in his third law of motion that "for every action there is an equal and opposite reaction." It is upon this principle that a rocket operates. Propellants (a high oxygen containing fuel or a mixture of fuel plus oxidant) are combined in a combustion chamber of the rocket where they chemically react to form a hot jet of gases (usually at a temperature of 3000°C and a pressure of 300 kg/cm^2) which are then accelerated and ejected at high velocity through a rocket nozzle, thereby imparting momentum to the engine. The thrust force of a rocket motor is the reaction experienced by the motor structure due to ejection of the high velocity matter. This is the same phenomenon which pushes a garden hose backward as water flows from the nozzle, or makes a gun recoil when fired

For a satellite to reach an orbit around Earth, the rocket must impart to the satellite a change in velocity of about 9.3–10 km/s consisting ~7.8 km/s for horizontal acceleration needed to reach orbital speed together with speed needed for gaining altitude and to compensate for gravity loss and atmospheric drag. The main proven technique involves launching nearly vertically for a few km while performing a gravity turn, and then progressively flattening the trajectory out at an altitude of 170+ km and accelerating on a horizontal trajectory (with the rocket angled upwards to fight gravity and maintain altitude) for a 5–8 minutes burn until orbital velocity is achieved. For a satellite to be in a stable orbit, 350 km is a more standard altitude for low Earth orbits otherwise the lower altitude will make satellites unstable due to atmospheric drag. For example, the international Space Station (ISS) orbits at an altitude of 330 km therefore it must be periodically re-boosted using its own engine or visiting spacecraft to maintain its orbit. Translunar and interplanetary flights are commonly launched into a temporary parking orbit, to allow spacecraft checkout, and more precise control of the final injection manoeuvre, rather than being launched directly to terminal velocity

1. Earth
1.2 First images of Earth from space

Image on the top: the first official image of Earth to be taken from space. Credit: White Sands Missile Range/Applied Physics Laboratory

The first image of Earth from space was taken on the sub-orbital V-2 rocket flight launched by the U.S. on October 24, 1946. Years before the world entered the space age, this grainy, black-and-white image was taken from an altitude of 105 km by a 35 millimeter motion picture camera

Image on the bottom: the first image panorama of Earth to be taken from space. Credit: White Sands Missile Range/Applied Physics Laboratory

Later in the same year, using the images taken by the same V-2 rocket, scientists, engineers and soldiers stitched the image frames together to create the first panorama of part of Earth covering millions of kilometres at a single glance against the blackness of space

1. Earth
1.3 First images of Earth from satellite

Image on the left: the first orbital satellite image of Earth was made on August 14, 1959 by the U.S. Explorer 6. Credit: NASA. The image shows a sun-lighted area of the Central Pacific ocean and its cloud cover

Artificial satellite is a manmade object which has been intentionally placed into orbit, and different from natural satellites such as Earth's Moon. Satellites are propelled by rockets to their orbits and are usually semi-independent computer-controlled systems. Digital satellite images are made from pixels, the smallest picture element of a digital image

Image on the right: the first image of Earth from a weather satellite, taken by the TIROS-1 satellite on April 1, 1960. Credit: NASA. This image from a weather satellite provided new information on cloud systems, including spiral formations associated with large storms, immediately proving their value to meteorologists. Parallel to military satellite missions, particularly reconnaissance over Soviet Union, TIROS-1 helped garner support for new satellite launches and continued operation of environmental satellites. So the era of 'remote sensing' has begun. Remote sensing is the science and art of identifying, observing, and measuring an object or phenomena without coming into direct contact with it. The first satellite image of the Earth might have been made by the Soviet Union, as Soviet satellite Luna 3 has photographed the far side of the Moon on October 6, 1959

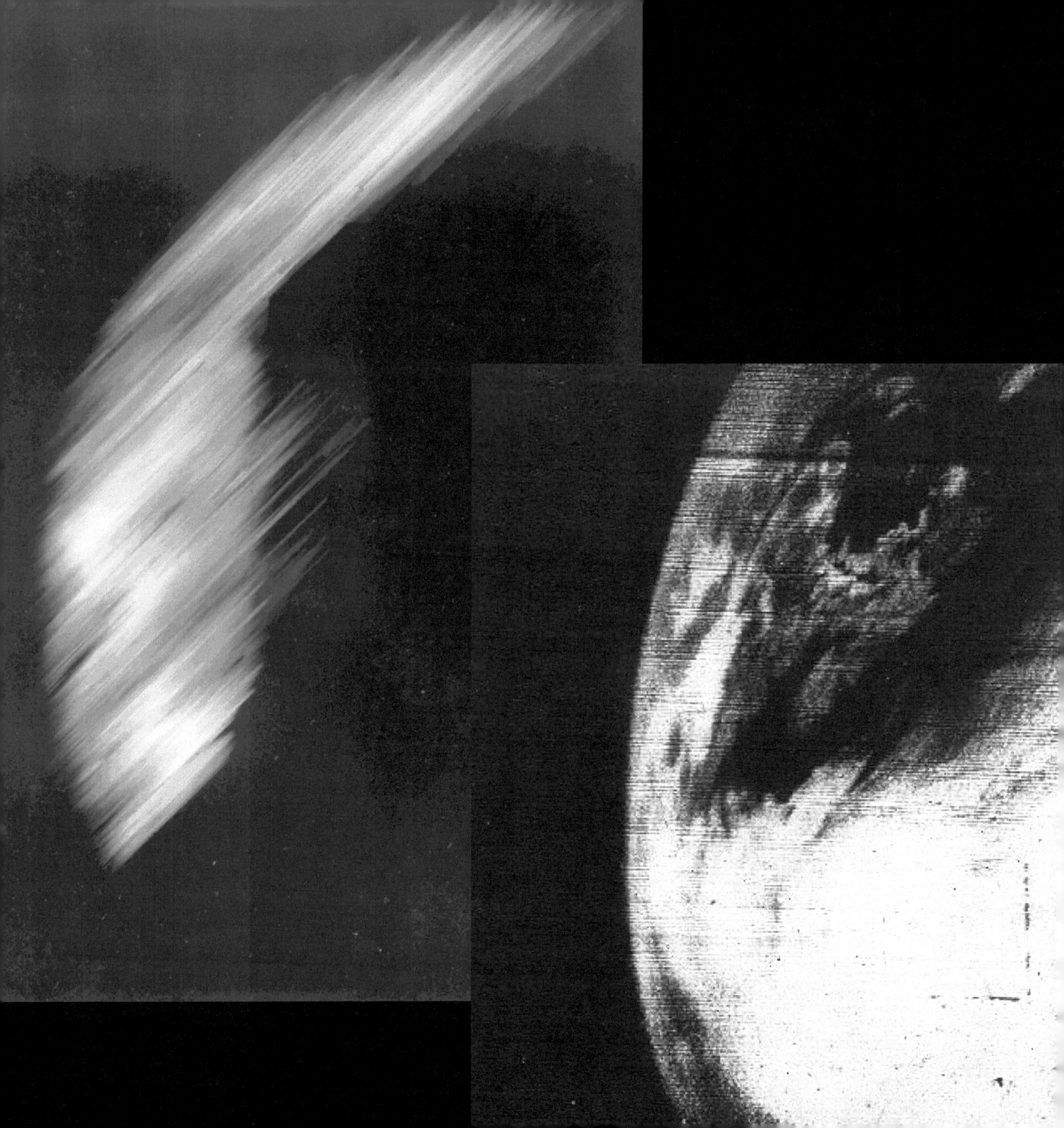

1. Earth
1.4 The first view of Earth from Moon

Lunar Orbiter 1 spacecraft image taken on August 23, 1966, restored in 2008 by NASA. Credit: NASA

This crescent of the Earth was photographed when the spacecraft was on its 16th orbit and just about to pass behind the Moon

Images of Earth from space had been taken before, by rockets in the 1940s, and satellites in the 1950s and 1960s (see the previous pages). However, those images captured just parts of Earth, as opposed to a full-on view of the planet. This image forever changed how we see our home planet Earth, providing the first look of Earth as a different kind of place from this really foreign kind of desolate landscape what we call Moon

The image was taken as preparation for sending humans to the Moon, particularly to search for safe landing places for Apollo missions

1. Earth
1.5 Earthrise

This image is called "Earthrise", the first image that humans had taken whilst actually behind the camera on board the first manned mission to the Moon, Apollo 8 on December 24, 1968. Credit: NASA

Apollo 8 was launched on December 21, 1968, and became the first manned spacecraft to leave Earth orbit, reach the Earth's Moon, orbit it and return safely to Earth

Crew members called **astronauts**, Commander Frank Borman, Command Module Pilot Jim Lovell, and Lunar Module Pilot William Anders became the first humans to travel beyond low Earth orbit, the first to see Earth as a whole planet, the first to directly see the far side of the Moon, and then the first to witness Earthrise

They held a live broadcast from lunar orbit, in which they showed pictures of the Earth and Moon as seen from their spacecraft and famously said: "the vast loneliness is awe-inspiring and it makes you realize just what you have back there on Earth"

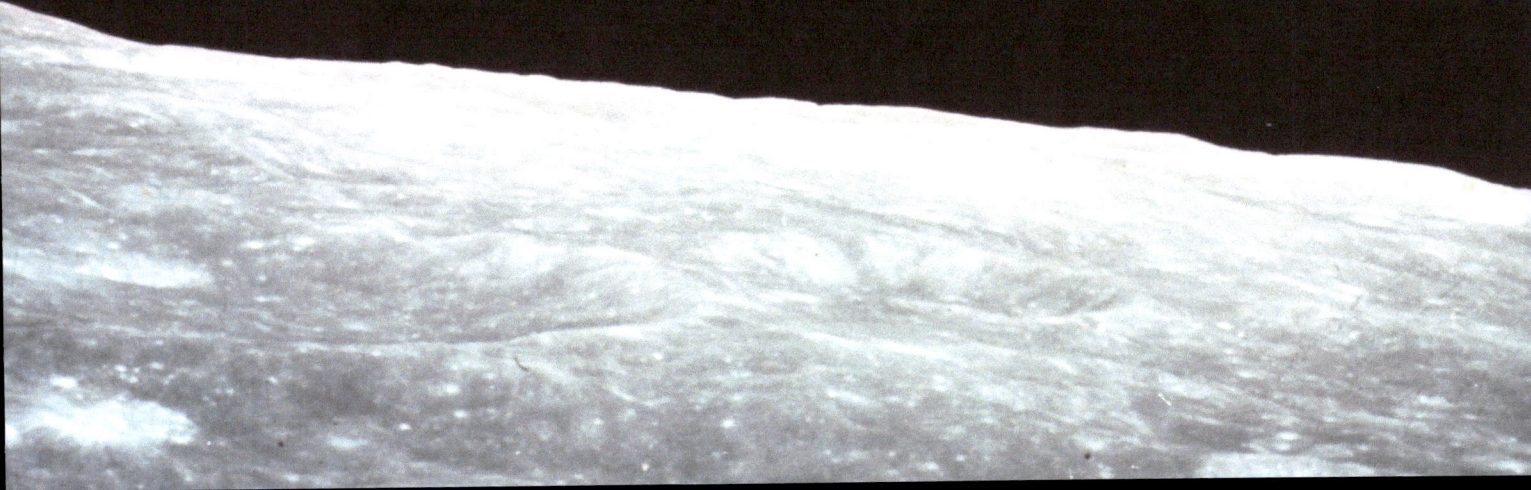

1. Earth
1.6 Blue Marble and first satellite maps

Top left: the "Blue Marble" is the first full-view image of the entire planet Earth, taken on December 7, 1972, by the crew of the Apollo 17 spacecraft, at a distance of about 45,000 km as they traveled back from the Moon with the Sun behind them illuminating our world. It is one of the most widely distributed images in human history.
Credit: NASA

Top right: the "Black marble" is Earth at night image captured by Suomi NPP satellite, composited of data acquired over 9 days in April 2012 and 13 days in October 2012. Credit: NASA

Bottom: the first Satellite Map of the Earth called GeoSphere which marked a milestone in cartographic history. The map was completed in April 15, 1990 showing the real world as it appears from space. The work required one year of effort by Van Sant and his technical team using the world's most powerful graphic computers at the time.
Credit: Tom Van Sant/GeoSphere Project

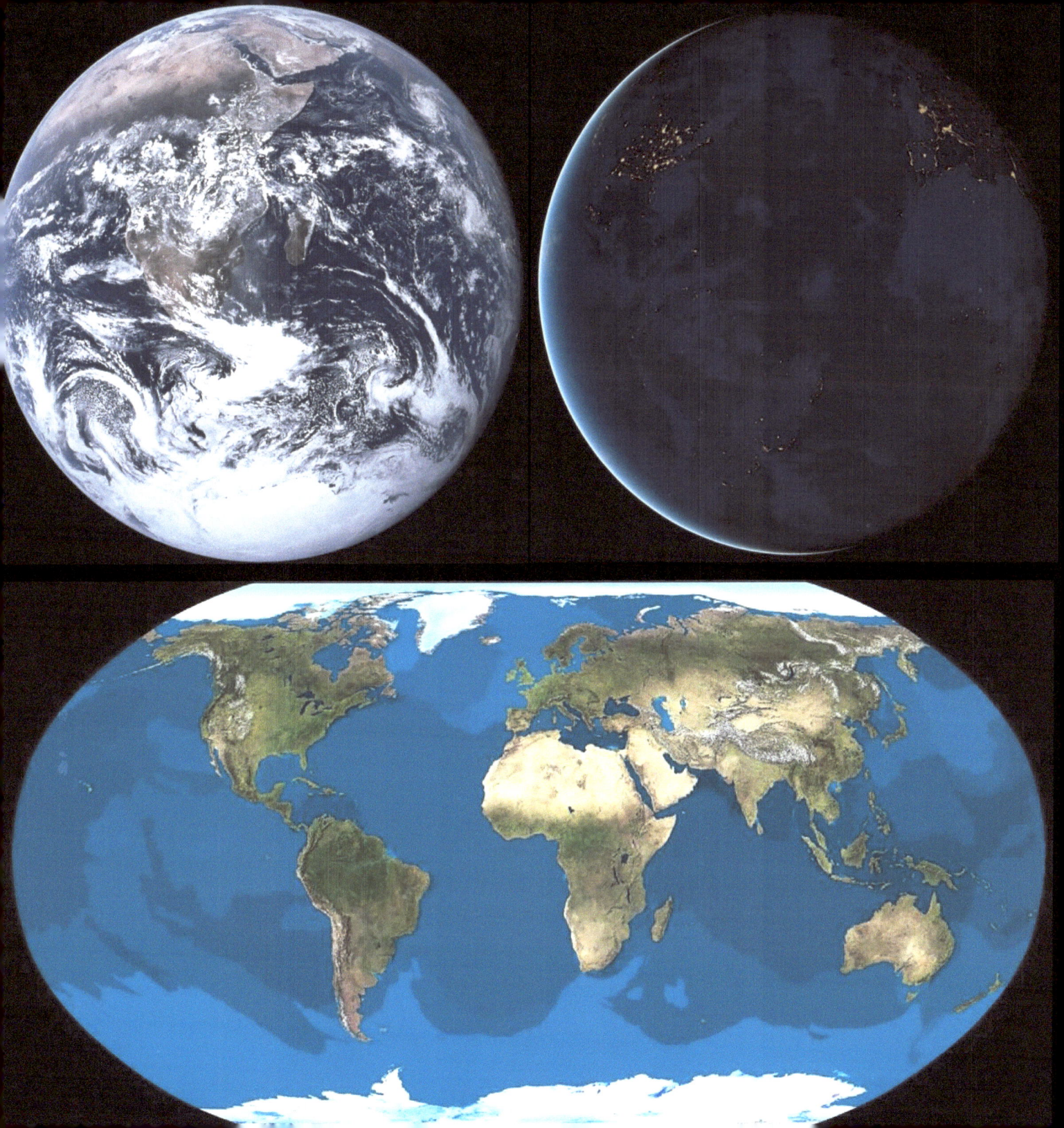

An image called "Pale Blue Dot" is a photograph of Earth taken on February 14, 1990, by the Voyager 1 space probe from a record distance of about 6 billion km, as part of the Family Portrait series of the Solar System. Earth appears as a tiny dot (the bluish-white speck approximately halfway down the brown band to the right) within the darkness of deep space. The famous astrophysicist Carl Sagan said the following about the Pale Blue Dot: "That's us. On it everyone you love, everyone you know, everyone you ever heard of, every human being who ever was, lived out their lives". Credit: NASA

The Cassini spacecraft has captured Saturn's rings and our planet Earth and its Moon in the same frame from 1.44 billion km away on July 19, 2013. In the image Earth appears as a blue dot at centre right; the Moon can be seen as a fainter protrusion off its right side. Credit: NASA

In its long journey to Jupiter, Juno spacecraft captured Earth with its Moon from 10 million km from Earth on August 26, 2011. The two dots, the Earth and Moon, are about 402,000 km apart. Credit: NASA

This is what Earth looks like from the Moon surface as photographed by Commander Eugene Cernan, a crew member of the Apollo 17 mission in December, 1972. Unlike the "Earthrise" images which were taken from the Moon orbiting spacecrafts, this image is taken from the Moon surface therefore shows what Earth crescent would look like for a person standing next to the rock on Moon surface. Credit: NASA Johnson/ Eugene Cernan/ Kipp Teague/Glen Swanson

2. Our planets
2.1 A brief background of our planets

Our planets include all eight planets that orbit the Sun. Order of the planets by the distance from the Sun: Mercury, Venus, Earth, Mars, Jupiter, Saturn, Uranus, and Neptune. Another large body is Pluto, now classified as a dwarf planet or plutoid. A planet is an astronomical object orbiting a star or stellar remnant that: is massive enough to be rounded by its own gravity; is not massive enough to cause thermonuclear fusion; and has cleared its neighbouring region of planetesimals

Our planets and moons which can be seen by naked eye in order of their maximum brightness: Moon, Venus, Jupiter, Mars, Mercury and Saturn

The inner planets closest to the Sun, i.e. Mercury, Venus, Earth, and Mars are terrestrial planets. Terrestrial planet (Earth-like or rocky planet) is a planet that is composed primarily of silicate rocks or metals. Terrestrial planets have a solid planetary surface, making them substantially different from the larger giant planets. Our Solar System has four giant planets, i.e. Jupiter, Saturn, Uranus and Neptune, which are composed mostly of some combination of low-boiling-point materials (gases or ices). Jupiter and Saturn are gas giants, whereas Uranus and Neptune are ice giants

Fact Box 2: true and false colour images

True colour images are recorded in blue, green and red wavelengths and rendered in natural colour for human eyes. This means that the colours of an object in an image or computer screen appear to a human observer the same way as if this observer were to directly view the object. For example, a green tree appears green in the image, a red apple red, a blue sky blue, and so on. False colour images are recorded in visible and outside visible wavelengths and rendered in natural colour spaces in blue, green and red channels to ease the detection of features that are not readily discernible by human eyes otherwise. The choice of different wavelengths to render in natural colour space (blue, green and red) is governed by the physical properties of the object under observation

The image below shows a true colour observation over the Great Rift Valley of Ethiopia taken by the American Landsat 8 satellite (Credit: NASA/USGS). We see rift valley lakes with different colour pigments, clouds, towns and vegetation in green area the same way if we were looking the scene from above. The pixel size is 30 metres and the image covers hundreds of km

Observation from the same data acquisition on the left, but here rendered in false colour space to detect vegetation. In this false colour image, blue is rendered in blue, green is rendered in green, and infrared is rendered in red channel. Higher red tone shows higher vegetation density. Green leaves reflect half of the infrared radiation they intercept, therefore infrared is a useful wavelength to study vegetation density

Observation from the same data acquisition, but here rendered in false colour space to detect water bodies. In this false colour image, far infrared is rendered in blue, mid infrared is rendered in green, and near infrared is rendered in red channel. Water absorbs most of infrared radiation therefore it appears black on the image despite having different pigments in the true colour space. Clouds appear white in all colour spaces because they reflect almost all of radiations in the rendered wavelengths

2. Our planets
2.2 *Mercury*

Mercury is the smallest, the first from the Sun and the innermost planet in the Solar System. The diameter of Mercury is about 4880 km and it has no known natural satellites (Moon)

Partly because it has almost no atmosphere to retain heat, Mercury's surface temperature varies diurnally more than any other planet in the Solar System, ranging from −173 °C (−280 °F) at night to 427 °C (800 °F) during the day in some equatorial regions

Mercury's surface is heavily cratered and similar in appearance to the Moon, indicating that it has been geologically inactive for billions of years

Because Mercury orbits the Sun within Earth's orbit (as does Venus), it can appear in Earth's sky in the morning or the evening, but not in the middle of the night

False colour full view of mercury (top left), oblique view of Mercury's surface (top right), and a mosaic of oblique images of Abedin crater (bottom) captured by MESSENGER spacecraft orbiting Mercury between 2011 and 2015. Credit: NASA/Johns Hopkins University Applied Physics Laboratory/Carnegie Institution of Washington

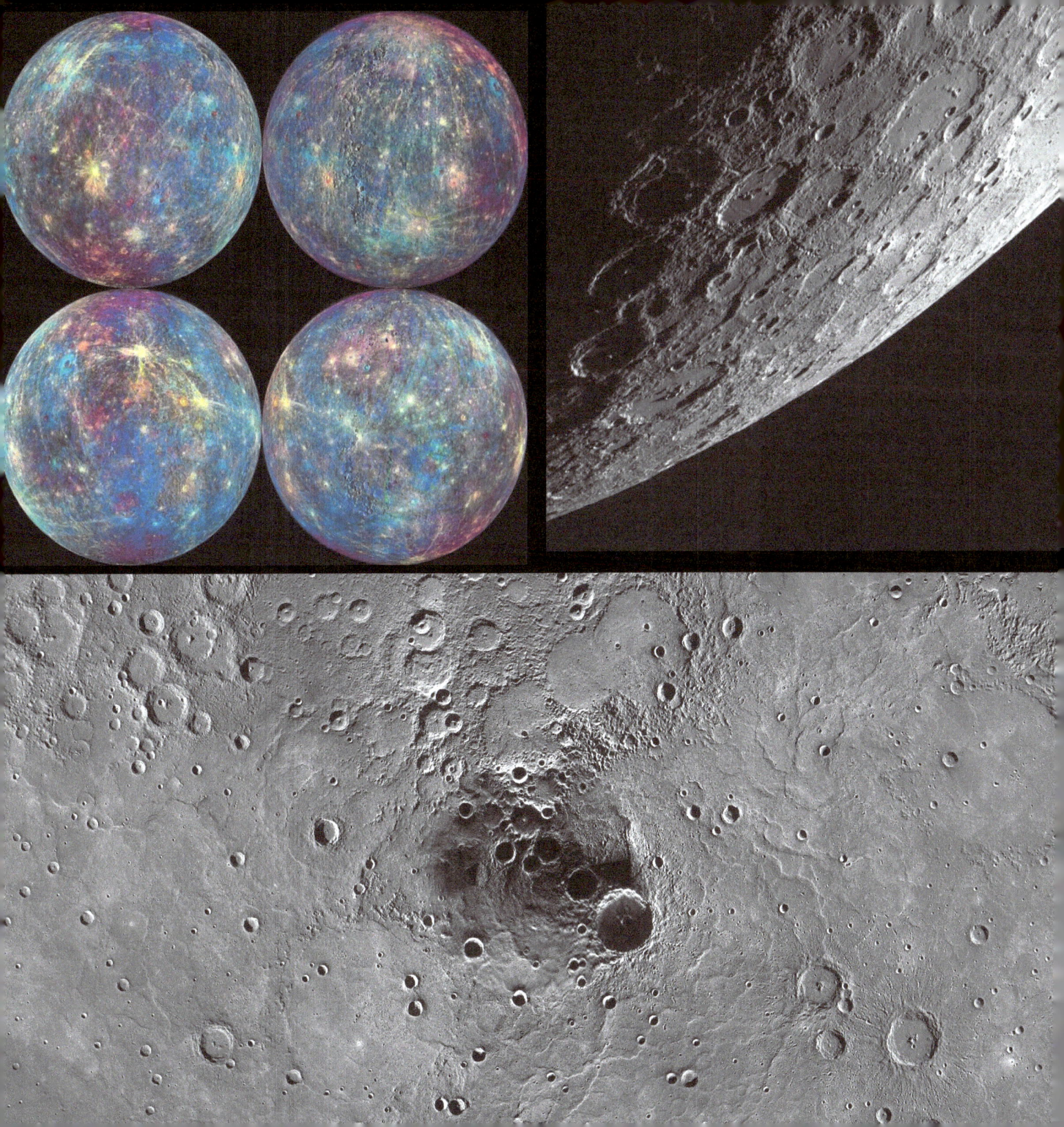

2. Our planets
2.3 Venus

Venus is the second planet from the Sun. The diameter of Venus is 12,092 km. It has no known natural satellites (Moon). Venus is by far the hottest planet in the Solar System, with a mean surface temperature of 462 °C (863 °F), even though Mercury is closer to the Sun. Venus has dense atmosphere consisting of more than 96% carbon dioxide and its thick clouds of sulfuric acid render observation of its surface impossible in visible light. Sunlight that passes through the atmosphere to heat the surface is trapped by the dense atmosphere and does not escape into space. This makes Venus hotter than Mercury

As second-brightest natural object in the night sky after the Moon, Venus has been part of human culture as far back as few thousand years usually described as "morning star" and "evening star"

Top panel: the crude color images of Venus surface from Venera 13 Lander, March, 1982. Credit: NASA

Middle panel: a real photograph from ESO's La Silla Observatory in northern Chile Jupiter (top), Venus (lower left, the brightest), and Mercury (lower right). Credit: ESO

Lower panel: global cloud free radar image of Venus from Magellan spacecraft compiled during 1990–1994 (left- Image Credit: NASA/JPL/USGS); ultraviolet image of clouds of Venus as seen by the Pioneer Venus Orbiter on February 26, 1979 (middle- Image Credit: NASA/ NSSDCA); and true colour of Venus from MESSENGER spacecraft flyby on Jane 5, 2007 (right- Image Credit: NASA/Johns Hopkins University Applied Physics Laboratory/Carnegie Institution of Washington)

2. Our planets
2.4 *Earth*

Earth is the third planet from the Sun, the densest planet in the Solar System, the largest of the Solar System's four terrestrial planets, and the only astronomical object known to harbor life. See chapter 1 for details of earth

Earth's only permanent natural satellite, the Moon, by its gravitational relationship with Earth, causes ocean tides, stabilizes the orientation of Earth's rotational axis, and gradually slows Earth's rotational rate

The Moon is thought to have formed approximately 4.5 billion years ago, not long after Earth

Since the rotational period of the Moon is exactly the same as the orbital period around the Earth, the same physical half of the Moon's sphere is always facing the Earth. The visible portion is called "near site" while the remaining half called "dark side" of the Moon which is never visible from Earth

There have been only six manned landings on Moon, between 1969 and 1972, all by U.S., a total of twelve men (astronauts) have landed on the Moon, two per mission (Apollo 11-17)

Photographs from successfully landed manned missions to Moon starting from panel 1 Apollo 11 (July 1969), Apollo 12 (November 1969), Apollo 14 (February 1971), Apollo 15 (August 1971), Apollo 16 (April 1972), and panel 6 is Apollo 17 (December 1972) mission. Credit: NASA Johnson/ Kipp Teague/Eric M. Jones/Ken Glover/Glen Swanson

An image of the Chinese Yutu rover taken by the Chang'e 3 lander unit shortly after touchdown on the Moon in December 2013. Credit: Chinese Academy of Sciences/China National Space Administration/The Science and Application Center for Moon and Deepspace Exploration/Emily Lakdawalla

An image of a rock formation on the Moon, named Long Yan (Pyramid Rock) captured by the Chinese Yutu rover on January 13, 2014. Credit: Chinese Academy of Sciences/China National Space Administration/The Science and Application Center for Moon and Deepspace Exploration/Emily Lakdawalla

Part of a full 360-degree panorama taken by the Chinese Chang'e 3 lander on December 17, 2013, three days after landing on the Moon. To the left is a crater nearly 20 meters in diameter whose rim features large blocks of bright rock. Credit: Chinese Academy of Sciences/China National Space Administration/The Science and Application Center for Moon and Deepspace Exploration/Emily Lakdawalla

2. Our planets
2.5 Mars

Mars is the fourth planet from the Sun and the second-smallest planet in the Solar System, after Mercury. Mars is often referred to as the "Red Planet" due to the iron oxide (rust) prevalent on its surface giving it a reddish appearance

Mars has two small natural satellites (Moons), Phobos (about 22 km in diameter) and Deimos (about 12 km in diameter). Mars is approximately half the diameter of Earth. Mars can easily be seen from Earth with the naked eye, as can its reddish coloring

Latest studies suggest that Mars has water albeit attached to other compounds or in the form of ice or briny flowing salt water. The two polar ice caps appear to be made largely of water but liquid water cannot exist on the surface of Mars due to low atmospheric pressure

The rotational period and seasonal cycles of Mars are likewise similar to those of Earth, as is the tilt that produces the seasons

Martian surface temperatures varies from −143 °C (−225 °F) at the winter polar caps to 35 °C (95 °F) in equatorial summer

1st panel: the first two from Hubble telescope (Credit: NASA/ESA/the Hubble Heritage Team (STScI/AURA)/J. Bell (ASU)/M. Wolff (Space Science Institute)) and the last two colour image of Phobes and Deimos captured by the Mars Reconnaissance Orbiter on 23 March 2008 (Credit: NASA/JPL-caltech/University of Arizona)

2nd panel: from left to right, false colour image of Columbia Hills, the 1st true colour image taken of another planet, false colour image of a rock called Mimi, and true colour image of a rock, all taken by Spirit rover between 2005-2006. Credit: NASA/JPL/Cornell

3rd – 5th panels: true colour images from Curiosity rover. Credit: NASA/JPL-Caltech/MSSS

Mars as viewed by NASA's Curiosity rover. Credit: NASA/JPL

Dark Sand Cascades on Mars

They might look like trees on Mars, but they're not. Groups of dark brown streaks have been photographed by the Mars Reconnaissance Orbiter on melting pinkish sand dunes covered with light frost. The above image was taken in April 2008 near the North Pole of Mars. At that time, dark sand on the interior of Martian sand dunes became more and more visible as the spring Sun melted the lighter carbon dioxide ice. When occurring near the top of a dune, dark sand may cascade down the dune leaving dark surface streaks -- streaks that might appear at first to be trees standing in front of the lighter regions, but cast no shadows. Objects about 25 centimeters across are resolved on this image spanning about one km. Close ups of some parts of this image show billowing plumes indicating that the sand slides were occurring even while the image was being taken. Credit: NASA/JPL/University of Arizona

2. Our planets
2.6 Jupiter

Jupiter is the fifth planet from the Sun. With diameter of 142,984 km at its equator, it is the largest in the Solar System, with a mass one-thousandth that of the Sun, with radius one-tenth of the Sun, and mass of two and a half times that of all the other planets in the Solar System combined

As a gas giant, Jupiter lacks a well-defined solid surface and hence its appearance constantly changes. Jupiter has at least 67 Moons and three faint planetary ring systems

Jupiter's volume is that of about 1,321 Earths, but it is only 318 times as massive. Jupiter has the largest planetary atmosphere in the Solar System, spanning over 5,000 km in altitude. Jupiter radiates more energy into space than it receives from the Sun. Jupiter's surface gravity is the strongest in our Solar System reaching at surface about 24.79 m/s^2

The best known feature of Jupiter is the Great Red Spot, permanent storm that is upto 3 times larger than Earth (see the red spot on the images)

1st panel: Jupiter, its Moons, aurora, and Great Red Spot from Hubble telescope. Credit: NASA/ESA/J. Nichols/Hubble Heritage Team (STScI/AURA)/A. Simon/M. Wong/G. Orton

2nd panel: Jupiter, its Moons and Great Red Spot from Cassini spacecraft flyby: Credit: NASA/JPL-Caltech/SSI/GSFC/ SA/JPL/University of Arizona/Space Science Institute

3rd panel: Jupiter's Moon Europa. 4th panel: the four largest Moons of Jupiter which are known as the Galilean satellites. From left to right Io, Europa, Ganymede, and Callisto. All from Galileo spacecraft. Credit: NASA/JPL-Caltech/SETI Institute

5th panel: Juno spacecraft true colour image after entering Jupiter's orbit. Some of Jupiter's Moons are visible. Credit: NASA/JPL-Caltech/SwRI/MSSS

2. Our planets
2.7 Saturn

Saturn is the sixth planet from the Sun and the second-largest in the Solar System, after Jupiter. The best known feature of Saturn is its prominent nine continuous main rings and three discontinuous arcs which are composed mostly of ice particles with a smaller amount of rocky debris and dust. The rings extend from 6,630 to 120,700 km outward from Saturn's equator, averaging approximately 20 meters in thickness

Sixty-two Moons and hundreds of Moonlets comprising the rings are known to orbit Saturn. Titan, Saturn's largest Moon, and the second-largest in the Solar System, is larger than the planet Mercury, and is the only Moon in the Solar System to have a substantial atmosphere and the only object other than Earth where clear evidence of stable bodies of surface liquid has been found (probably of liquid methane and ethane)

As a gas giant, Saturn lacks a well-defined solid surface, though it may have a solid core. The outermost layer spans 1,000 km and consists of gas. Together, Jupiter and Saturn hold 92% of the total planetary mass in the Solar System. Saturn radiates 2.5 times more energy into space than it receives from the Sun

1st and 2nd panels: true colour images of Saturn, its rings and huge storm, full Sun eclipse from Cassini probe. Credit: NASA/JPL/Space Science Institute
3rd panel: Titan, the largest Saturn Moon, and its atmosphere, surface and hydrocarbon lakes from Cassini-Huygens probe. Credit: NASA/JPL-Caltech/University of Arizona/University of Idaho/ESA/USGS

2. Our planets
2.8 *Uranus*

Uranus is the seventh planet from the Sun, and it has the third largest planetary radius and the fourth largest planetary mass in the Solar System. As ice giant, Uranus's atmosphere contains hydrogen, helium, and ices of water, ammonia, and methane, along with traces of other hydrocarbons. Uranus has the coldest planetary atmosphere in the Solar System, with a minimum temperature of −224.2 °C (-371.56 °F)

Uranus's mass is roughly 14.5 times that of Earth, making it the least massive of the giant planets

The extreme pressure and temperature deep within Uranus may break up the methane molecules, with the carbon atoms condensing into crystals of diamond that rain down through the mantle like hailstones

Uranus has thirteen known distinct rings which are very dark, faint and narrow. Uranus does not have a well defined solid surface

Uranus has 27 known Moons

Uranus is visible to the naked eye in dark skies

1st panel: images of Uranus and its rings. 2nd panel: Icy Moons of Uranus from Voyager 2 probe. Credit: NASA/JPL-Caltech

3rd panel: clouds, rings and Moons of Uranus from Hubble telescope. Credit: NASA/JPL/STScI/Erich Karkoschka/University of Arizona/ESA

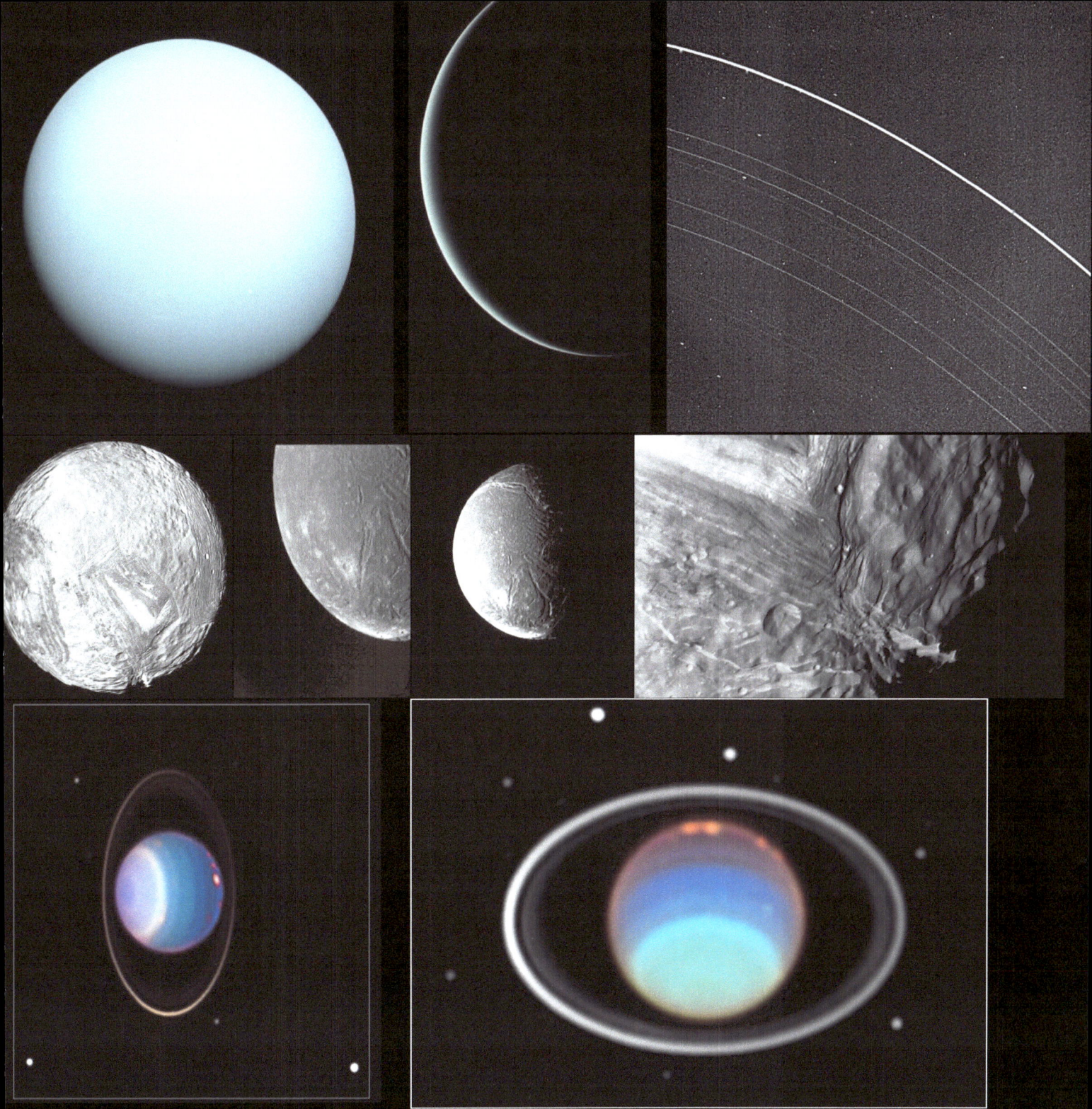

2. Our planets
2.9 Neptune

Neptune is the eighth and farthest known planet from the Sun in the Solar System. With 17 times the mass of Earth, Neptune is the fourth-largest planet by diameter, the third-most-massive planet, and the densest giant planet. Neptune is not visible to the naked eye and is the only planet in our Solar System found by mathematical prediction by Urbain Le Verrier rather than by empirical observation. Neptune has 14 known Moons. As ice giant, Neptune is similar in composition to Uranus. Neptune does not have a well defined solid surface

Because of its distance from Sun, Neptune's atmosphere is in average the coldest in our Solar System, approaching −218 °C (−360 °F), although Uranus's minimum temperature is colder than that of Neptune's. Like the other 3 giants, Neptune has a faint and fragmented ring system

Neptune is characterised by the strongest storms in our Solar System with wind speed nearly reaching supersonic flow — 600 m/s (2,200 km/h or 1,300 m/h)

1st panel: images of Neptune, its atmosphere, Great Dark Spot, and rings from Voyager 2 probe. Credit: NASA/JPL. 2nd panel: images of Triton, the largest Moon of Neptune from Voyager 2 probe. Credit: NASA/JPL

3rd panel: four global images of Neptune taken at four hours intervals by Hubble Space Telescope on June 25-26, 2011 to commemorate its first orbit around the Sun since its discovery. Neptune travels along in its huge orbit, slowly completing one revolution around the Sun approximately every 165 years. Credit: NASA/ESA/Hubble Heritage Team (STScI/AURA)

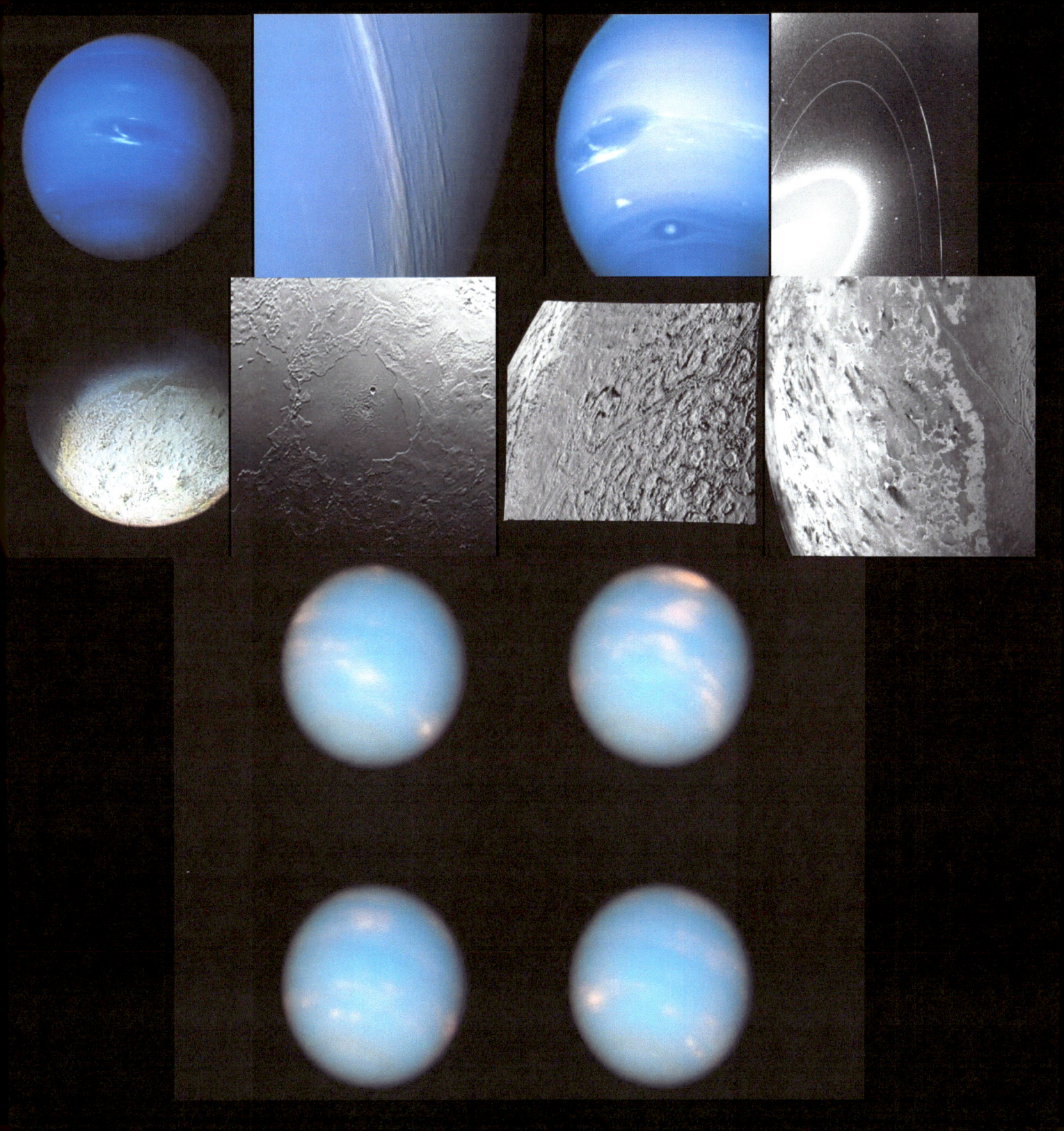

2. Our planets
2.10 Pluto

Pluto is a dwarf planet or plutoid in Kuiper belt. Between its discovery in 1930 and discovery of several objects of similar or even larger size in the Kuiper belt in 2005, Pluto was considered as the ninth planet

Pluto is the largest and second-most-massive known dwarf planet in the Solar System and the ninth-largest and tenth-most-massive known object directly orbiting the Sun. It is the largest known trans-Neptunian object by volume but is less massive than Eris, another dwarf planet which is the ninth-most-massive known body in the Solar System

Pluto is about one-sixth the mass of our Moon and one-third its volume and like other Kuiper belt objects, it is primarily made of ice and rock

Light from the Sun takes about 5.5 hours to reach Pluto while 8.33 minutes to the Earth. NASA has designated "Pluto time" to show how weak the amount of Sun light is on Pluto. Usually Pluto time is equivalent to twilight on Earth, about 4 minutes after sunset

Pluto's diameter is 2378 km. Its surface area is roughly the same surface area of Russia. Pluto has five known Moons while Charon, the largest of all with a diameter just over half that of Pluto, makes Pluto and Charon binary dwarf planets (not confirmed yet)

1st panel: global view of Pluto from New Horizons probe taken on 25 July 2015. See next pages for global and surface views of Pluto from New Horizons best images. 2nd panel: Jagged Ice shorelines and snowy pits, strange terrain, and vistas of Pluto. 3rd panel: Pluto's largest Moon Charon, and Charon and Pluto in true scale. All images are from New Horizons probe. Credit: NASA / Johns Hopkins University Applied Physics Laboratory / Southwest Research Institute

Global views of Pluto from New Horizons probe taken in July 2015. Credit: NASA / Johns Hopkins University Applied Physics Laboratory / Southwest Research Institute

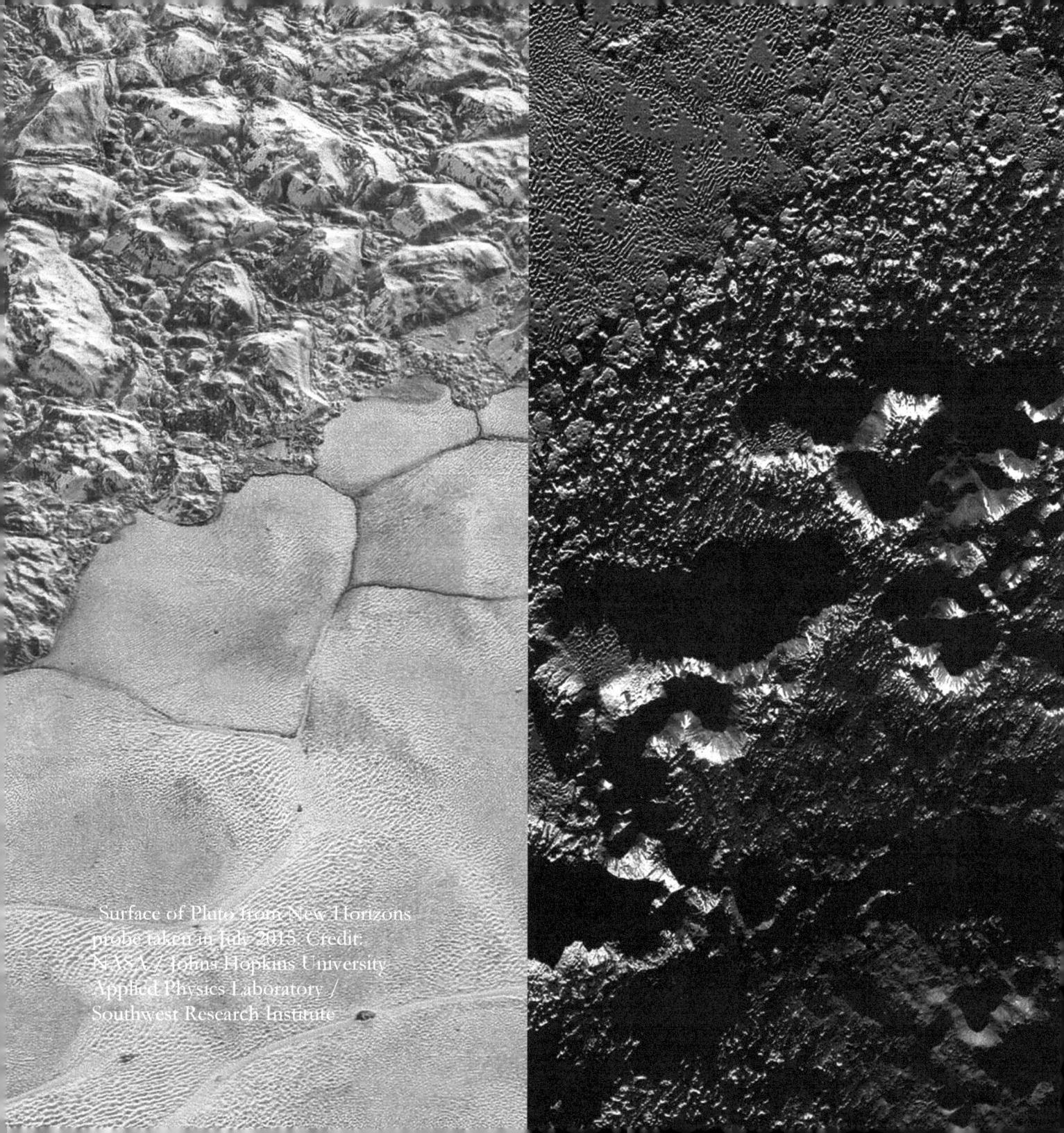

Surface of Pluto from New Horizons probe taken in July 2015. Credit: NASA / Johns Hopkins University Applied Physics Laboratory / Southwest Research Institute

2. Our planets
2.11 Other notable object systems in our Solar System

1. **The main asteroid belt:** is the circumstellar disc in the Solar System, occupied by numerous irregularly shaped bodies called asteroids or minor planets, located roughly between the orbits of the planets Mars and Jupiter. About half the mass of the main asteroid belt is contained in the four largest asteroids: Ceres, Vesta, Pallas, and Hygiea. Ceres is the only dwarf planet in the asteroid belt while the remaining bodies range down to the size of a dust particle. 99.8 percent of meteorites are believed to have originated in the asteroid belt. 1st panel: an image of asteroid called Gaspra taken by Galileo spacecraft on October 29, 1991. Credit: USGS/NASA/JPL. 2nd panel: global and surface images of Ceres taken by Dawn probe on May, 2015. Credits: NASA / JPL-Caltech / UCLA / MPS / DLR / IDA / Justin Cowart

2. **The Kuiper belt:** is a circumstellar disc in the Solar System extending beyond the orbit of Neptune. It is similar to the asteroid belt, but it is far larger—20 times as wide and 20 to 200 times as massive, and most of its objects are composed largely of frozen volatiles, such as methane, ammonia and water. The Kuiper belt is home to three officially recognized dwarf planets: Pluto, Haumea, and Makemake. Pluto is the largest and most-massive member of the Kuiper belt. The objects within the Kuiper belt, together with the members of the scattered disc and any potential Hills cloud or Oort cloud objects, are collectively referred to as trans-Neptunian objects

3. Other classes of small Solar System objects in other regions include the near-Earth objects, the centaurs (between Jupiter and Neptune), the scattered disc objects (extends from just outside Uranus' orbit, out to thirty billion km from the Sun), the sednoids (trans-Neptunian object), and the Oort cloud objects (extends from about sixteen billion to three trillion km out from the Sun)

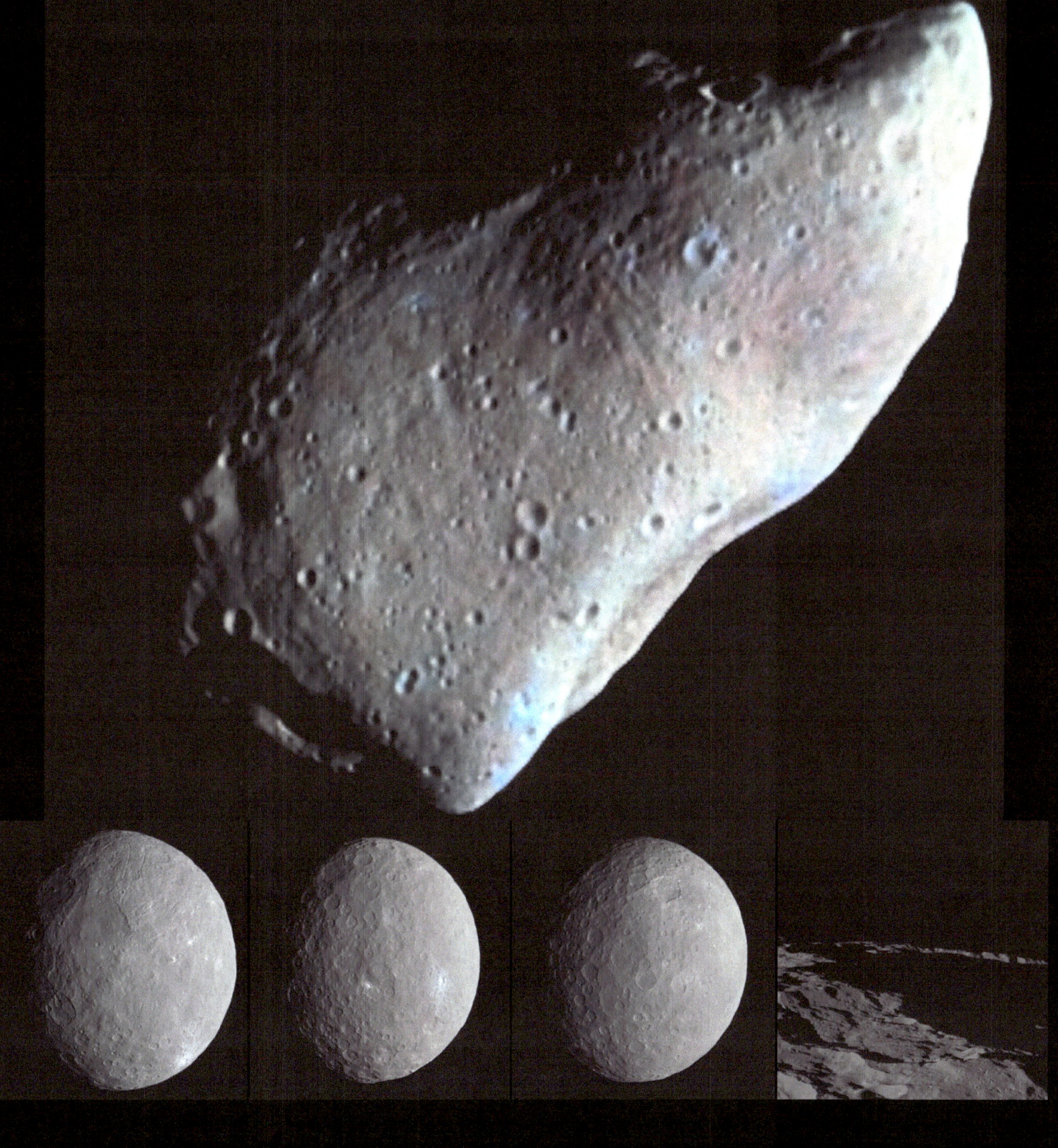

Image of comet 67P/Churyumov–Gerasimenko taken by Rosetta's navigation camera on 19 September 2014 at 28.6 km from the centre of the comet. Rosetta was a space probe built by the European Space Agency launched on 2 March 2004. Along with Philae, its lander module, Rosetta performed a detailed study of comet 67P/Churyumov–Gerasimenko (67P). Credit: ESA/Rosetta/NAVCAM

This image, confirming Philae's touchdown on the surface of Comet 67P/Churyumov–Gerasimenko arrived on Earth at 16:03 GMT on 12 November 2014. This is the first-ever soft landing on a comet by any space mission. Rosetta's lander Philae is safely on the surface of Comet 67P/Churyumov-Gerasimenko. One of the lander's three feet can be seen in the foreground. Credit: ESA/Rosetta//Philae/CIVA

3. Planetary system

A planetary system is a set of gravitationally bound non-stellar (stellar is a star) objects in orbit around a star or star system. Our planetary system called Solar System is the gravitationally bound system comprising the Sun and the objects that orbit it. Sun is the star of our planetary system

A star is a luminous sphere of plasma held together by its own gravity. For at least a portion of its life, a star shines due to thermonuclear fusion of hydrogen into helium in its core, releasing energy that traverses the star's interior and then radiates into outer space. Sceintists can determine the mass, age, metallicity (chemical composition), and many other properties of a star by observing its motion through space, its luminosity, and spectrum respectively. The total mass of a star is the main factor that determines its evolution and eventual fate

Stars are given a single-letter classification according to their spectra, ranging from type O, which are very hot, to M, which are so cool that molecules may form in their atmospheres. The Sun is a main sequence G yellow dwarf of intermediate temperature and ordinary size. Stars spend about 90% of their existence fusing hydrogen into helium in high-temperature and high-pressure reactions near the core. Such stars are said to be on the main sequence, and are called dwarf stars

The more massive the star, the shorter its lifespan, primarily because massive stars have greater pressure on their cores, causing them to burn hydrogen more rapidly. The most massive stars last an average of a few million years, while stars of minimum mass such as red dwarfs burn their fuel very slowly and can last tens to hundreds of billions of years. Stars range in size from neutron stars, which vary anywhere from 20 to 40 km in diameter, to supergiants like Betelgeuse in the Orion constellation, which has a diameter approximately 1,070 times that of the Sun

Fact Box 3: Observing Earth and Space

In remote sensing, consisting of Earth and space observations, including observational astronomy, the main source of information about terrestrial and celestial bodies is the electromagnetic spectrum. You know more about the electromagnetic spectrum than you may think. The image below (modified from NASA's Imagine the Universe) shows where you might encounter each portion of the electromagnetic spectrum in your day-to-day life

Most instruments used for scientific studies record each electromagnetic radiation spectrum emitted or reflected simultaneously and store them separately in pixels. The width in wavelength of each recorded spectrum varies depending on the purpose. For example, a common hand held digital camera including smart phone cameras record blue, green and red wavelengths to produce photographs for human eyes. Unlike scientific instruments, the three wavelengths in ordinary camera are not well defined and therefore cannot be used for scientific studies. Scientific instruments record precisely defined regions of the electromagnetic spectrum ranging from a single to hundreds of distinct spectral regions simultaneously. For example, water ice reflects almost all of the electromagnetic radiation in visible wavelengths that is why it appears white while it absorbs most of radiation in infrared wavelengths around 1.5 and 2 micrometers. This way by studying the spectral curve (reflected radiation in each wavelength), we can identify whether or not there is water ice on Mars and measure the amount it

Electromagnetic radiation can be expressed in terms of energy (measured in electron volts), wavelength (measured in unit of length such as nanometres, meters), or frequency (measured in cycles per second, or Hertz). Each of these three quantities for describing electromagnetic radiation are related to each other in a precise mathematical way

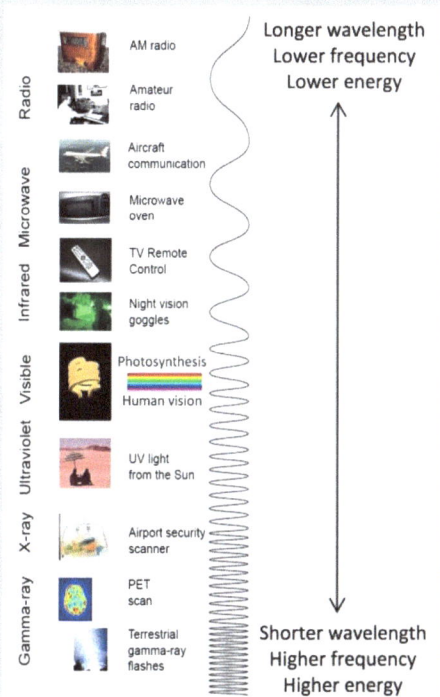

- Radio wavelength (wavelengths greater than approximately one millimetre) are used to study Earth's ice, supernovae, interstellar gas, pulsars, and active galactic nuclei

- Infrared wavelength (wavelengths longer than red light) are used to study Earth's plants, planets, and nebulae

- Visible wavelength (visible light consisting of blue, green and red lights) are used to study Earth's photosynthesis pigments but rarely used for space study

- Ultraviolet wavelength (wavelengths shorter than violet) are used to study stars

- X-ray wavelength are used to study X-ray binaries, pulsars, supernova remnants, elliptical galaxies, clusters of galaxies, and active galactic nuclei

- Gamma-ray wavelength (shortest wavelengths of the electromagnetic spectrum) are used to study pulsars, neutron stars, and black hole candidates such as active galactic nuclei

- In addition to electromagnetic radiation, scientists also use gravitational wave to collect observational data about compact objects such as black holes, and neutrino detection for cosmic ray (very high energy particles)

- Earth observation instruments also use electromagnetic spectrum ranging from visible light to radio wavelengths

Some parts of the electromagnetic spectrum can be observed from the Earth's surface, while other parts are only observable from outside the Earth's atmosphere

3. Planetary system
3.1 Solar System

The Solar System is consisted of our star what we call the Sun, the planets Mercury, Venus, Earth, Mars, Jupiter, Saturn, Uranus, Neptune and Pluto, and the satellites of the planets, numerous comets, asteroids, and meteoroids; and the interplanetary medium and objects. The eight planets of the Solar System by decreasing size are Jupiter, Saturn, Uranus, Neptune, Earth, Venus, Mars and Mercury

The Solar System was formed 4.6 billion years ago from the gravitational collapse of a giant interstellar molecular cloud. The Solar System will remain roughly as we know it today until the hydrogen in the core of the Sun has been entirely converted to helium, which will occur roughly 5 billion years from now. The vast majority of our Solar System's mass is in the Sun, with most of the remaining mass contained in Jupiter. Our Solar System's mass is 1.0014 times the Sun mass (or our Sun consists 99.86% of the system's known mass). The Sun, which comprises nearly all the matter in our Solar System, is composed of roughly 98% hydrogen and helium. Jupiter and Saturn, which comprise nearly all the remaining matter, are also primarily composed of hydrogen and helium. All planets have almost circular orbits that lie within a nearly flat disc called the ecliptic

Within our Solar System, Earth is the only planet known to have current geological activity, the only place where life is known to exist, and it is the only planet where plate tectonics has been observed. The synodic rotation of Earth on its axis and its orbit around the Sun are the basis of the predominant calendars we use. Viewed from Earth, the apparent rotational period of the Sun at its equator is about 28 days

Images on the right: Solar System family portrait from Voyegaer 1 probe 6 billion km away from Earth. Left-to-right and top-to-bottom are Venus, Earth ("Pale Blue Dot"), Jupiter, Saturn, Uranus, Neptune. The images were taken through three color filters – violet, blue and green and recombined to produce the false color images. Credit: NASA/JPL

The solar wind, a stream of charged particles flowing outwards from the Sun, creates a bubble-like region in the interstellar medium known as the heliosphere. The heliosphere and planetary magnetic fields (for those planets that have them) partially shield the Solar System from high-energy interstellar particles called cosmic rays. All the planets and most other objects orbit the Sun in the same direction that the Sun is rotating (counter-clockwise, as viewed from above Earth's North Pole). Objects closer to the Sun (with smaller semi-major axes) travel more quickly because they are more affected by the Sun's gravity

Along with light, the Sun radiates a continuous stream of charged particles (a plasma) known as the solar wind. Earth's magnetic field stops its atmosphere from being stripped away by the solar wind. Venus and Mars do not have magnetic fields, and as a result the solar wind is causing their atmospheres to gradually bleed away into space. Coronal mass ejections and similar events blow a magnetic field and huge quantities of material from the surface of the Sun. Aurorae seen near the Earth's magnetic poles are the results of the interaction of this magnetic field and material, with Earth's magnetic field which funnels charged particles into Earth's upper atmosphere

Comets are small Solar System bodies, typically only a few km across, composed largely of volatile ices. When a comet enters the inner Solar System, its proximity to the Sun causes its icy surface to sublimate and ionise, creating a coma: a long tail of gas and dust often visible to the naked eye. Old comets that have had most of their volatiles driven out by solar warming are often categorised as **asteroids**. Larger chunks of rock that come from the asteroid belt located between the orbits of Mars and Jupiter are generally called asteroids. Space debris smaller than an asteroid are called **meteoroids** which are smaller than a km and frequently only millimeters in size. Most meteoroids that enter the Earth's atmosphere are so small that they vaporize completely and never reach the planet's surface but if they do enter Earth's atmosphere, they are called **meteorite**. The flash of light caused by the debris burning up as it passes through our atmosphere is called **meteor** or what we commonly call **a shooting star**

Images on the right: global view of Sun imaged using different wavelengths of extreme ultraviolet light by SDO spacecraft. The dark region is corolar hole where the magnetic field of the Sun is more open, emitting streams of high-speed solar wind. See next page for the resulting solar flares the size of many Earths. Credit: NASA/Solar Dynamics Observatory

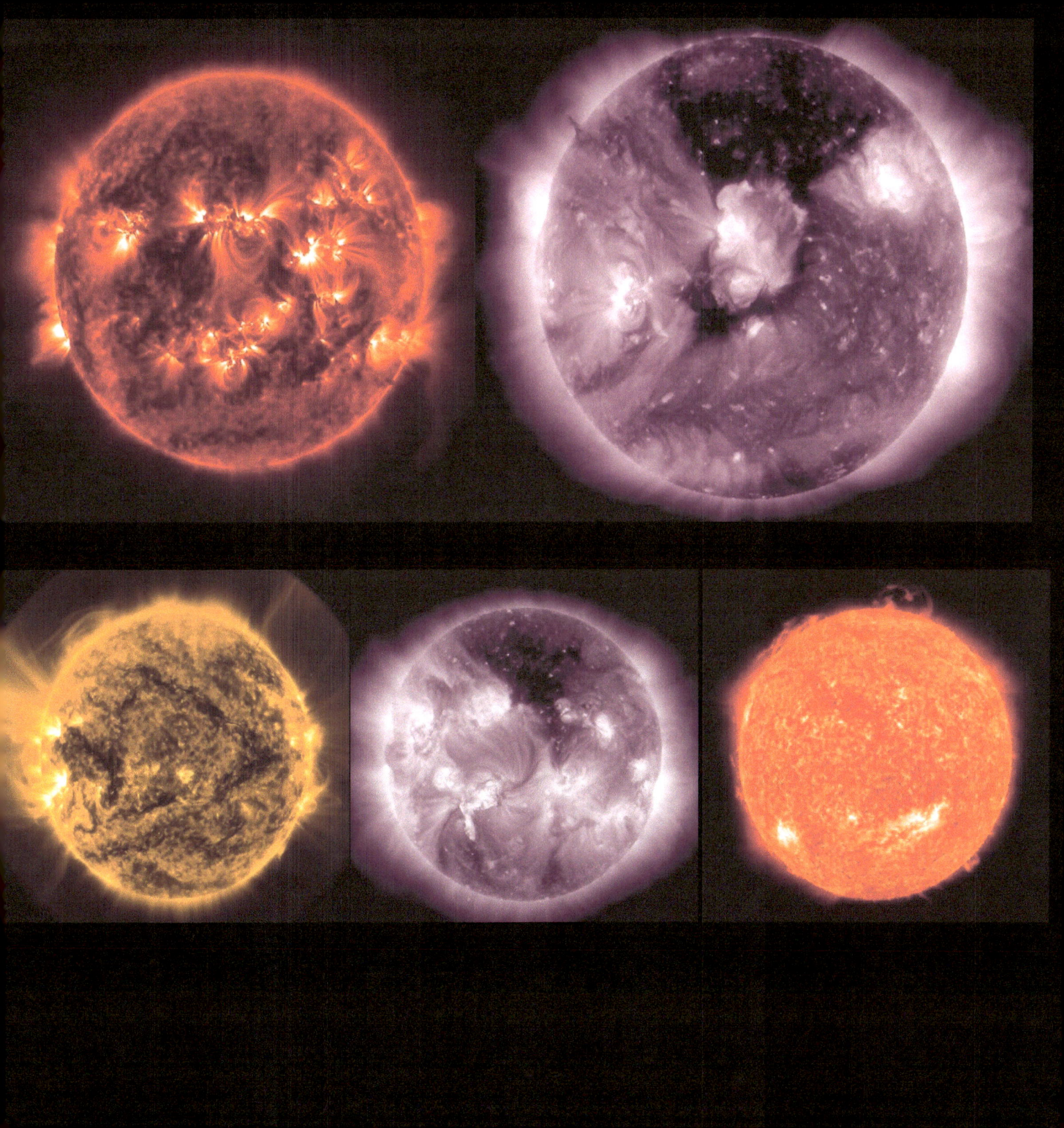

3. Planetary system
3.2 Sun

Sun is 332,900 Earth masses and produces temperatures and densities in its core high enough to sustain nuclear fusion of hydrogen into helium, which releases an enormous amount of energy, mostly radiated into space as electromagnetic radiation peaking in visible light. The energy of this sunlight supports almost all life on Earth by photosynthesis, and drives Earth's climate and weather

Although most people mentally picture the Sun as yellow, Sun's true color is white

The Sun is located in the Orion Arm, 26,000 light-years from the center of our galaxy called the Milky Way and its speed within the Milky Way is about 220 km/s, so that it completes one revolution every 225–250 million years. The Sun, and thus the Solar System, is found in what scientists call the galactic habitable zone, possibly produced by the supernova. It has been argued that the Sun's passage through the higher density spiral arms of our galaxy (see the spiral arms in the next chapter) often coincides with mass extinctions on Earth, perhaps due to increased impact events. Therefore, we expect mass extinction every 225–250 million years, the time it takes the Sun to orbit around the center of our galaxy

Sun's variation in the amount of energy emitted largely depends on the number and amount of sunspots. **Sunspots** are temporary phenomena on the surface of the Sun that appear as dark spots compared to surrounding regions and they are areas of reduced surface temperature caused by concentrations of magnetic field flux. Their number varies according to the approximately 11-year solar cycle, which results in changes in the amount of radiation from the sun. Higher sunspot numbers usually indicate higher radiation output from sun, intensified solar flares and auroras at moderate to high latitudes and the disruption of radio communications and electric power. For example, in the 17th century, few sunspots were observed for several decades and this coincided in time with the era of the Little Ice Age, when Europe experienced unusually cold temperatures. Larger sunspots are visible from Earth without telescope

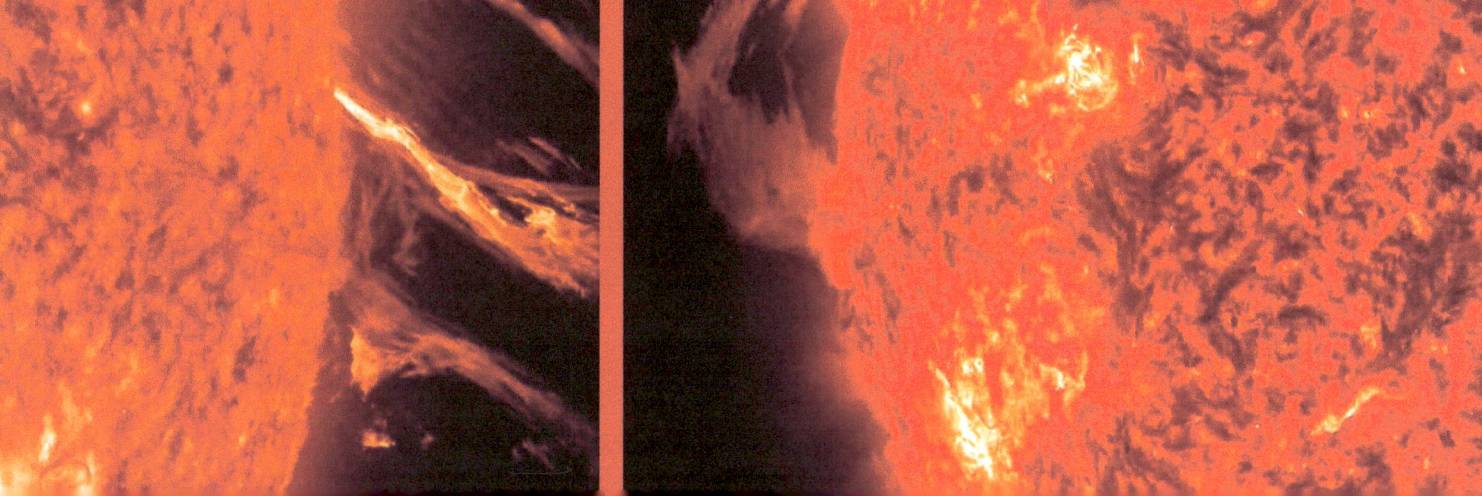

4. Galaxy
4.1 A brief background of Galaxy

A galaxy is a gravitationally bound system of stars, stellar remnants, interstellar gas, dust, and dark matter. Based on their visual morphology, there are three types of galaxies:

1. Elliptical galaxies - approximately ellipsoidal shape and a smooth, nearly featureless brightness profile, more three-dimensional, without much structure, and their stars are in somewhat random orbits around the center. They are spherical or ovoid masses of stars, starved of star-making gases. See 1st panel images on the right

2. Spiral galaxies - resemble spiraling pinwheels and consist of a flat, rotating disc containing stars, gas and dust, and a central concentration of stars known as the bulge. These are surrounded by a much fainter halo of stars, many of which reside in globular clusters. Our own galaxy called Milky Way has recently been confirmed to be a spiral galaxy, although the bar itself is difficult to observe from the Earth's current position within the galactic disc. See 2nd panel images on the right

3. Irregular galaxies - do not have a distinct regular shape, do not fall into any of the regular classes, and they are often chaotic in appearance, with neither a nuclear bulge nor any trace of spiral arm structure. See 3rd panel images on the right. Credit: NASA/ESA/The Hubble Heritage Team (STScI/AURA)

Galaxies range in size from dwarfs with just a few billion stars to giants with one hundred trillion stars, each orbiting its galaxy's center of mass, mostly a black hole

At the end of the stellar age, galaxies will be composed of compact objects: brown dwarfs, white dwarfs that are cooling or black dwarfs that are cold, neutron stars, and black holes (see next page)

4. Galaxy
4.2 Black hole, Neutron star and Pulsar

A black hole is a region of spacetime exhibiting such strong gravitational effects that nothing can escape from inside of it, including particles and electromagnetic radiation such as light. A black hole is usually formed when very massive stars collapse at the end of their life cycle, which can continue to grow by absorbing mass such as stars and other black holes from its surroundings. Black holes are invisible and their presence can be inferred through their interaction with other matter and with electromagnetic radiation such as visible lights. Only in the year 2016, the existence of black holes was indirectly confirmed observationally through gravitational waves generated from black holes merger and collisions. *Gravitational waves* are ripples in the curvature of spacetime that propagate as waves, generated in certain gravitational interactions that propagate outward from their source. The Milky Way's central black hole, known as Sagittarius A*, has a mass of four million times greater than the Sun

A neutron star is the collapsed core of a large star and is one of the smallest and densest celestial objects known to exist in the Universe with a radius of only about 20–40 km and a mass of about 1.4 times the Sun. Neutron stars are composed almost entirely of neutrons, which are subatomic particles with no net electrical charge. If the collapsed core of a large star has too great a mass, between 1.4 and 3 solar masses, it will continue to collapse into a black hole. As the star's core collapses, its rotation rate increases as a result of conservation of angular momentum, hence neutron stars rotate at up to several hundred times per second. As they do so, they can emit beams of electromagnetic radiation that makes them detectable as pulsars. A pulsar is a highly magnetized, rotating neutron star that emits a beam of electromagnetic radiation. This radiation can be observed only when the beam of emission is pointing toward the observer the same way as a lighthouse can be seen only when the light is pointed in the direction of an observer. Certain types of pulsars rival atomic clocks in their accuracy in keeping time. *Pulsars*, generally speaking, are rotating neutron stars and pulsars appear to pulse because they rotate

Images on the right are of the Crab Pulsar, taken by the Chandra X-ray Observatory over several months, show the bright white pulsar at the center, and jets of matter spewing away. Some pulsars emit X-rays. Credit: NASA/CXC/ASU/J.Hester et al.

360-degree panoramic image of the plane of our galaxy, Milky Way (Credit: ESO/S. Brunier)

If you look up into the night sky on a very clear night, in an area with very little light pollution, you will see a band of stars splashed across the sky. That band is a portion of the Milky Way. The Solar System is inside the disk of the Milky Way, therefore we cannot see our galaxy or its spiral structure from our planet because we are inside it and have no means of taking observations from above or below the galaxy. However, this 360-degree panoramic image from Earth covers the entire southern and northern celestial sphere with the plane of the Milky Way Galaxy cutting a luminous swath across the image

Hoag's Object: A Ring Galaxy (Credit: R. Lucas (STScI/AURA)/Hubble Heritage Team/ NASA)

This image is from NASA's Hubble Space Telescope captured on July 9, 2001. An image of Hoag's Object, a non-typical galaxy of the type known as a ring galaxy, discovered in 1950 by astronomer Art Hoag. A ring galaxy is thought to occur when a smaller galaxy passes through the core of a spiral galaxy. Serendipitously, from the perspective of our Solar System what appears to be an even more distant ring galaxy is plainly visible within the gap between this galaxy's central body of mostly yellow stars and the outer ring of blue stars

Merging pair of galaxies called the Antennae galaxies (Credit: ESA/(STScI/AURA)/Hubble collaboration/ NASA). This image from Hubble Space Telescope shows the two spiral galaxies started to interact a few hundred million years ago, making the Antennae galaxies one of the nearest and youngest examples of a pair of colliding galaxies. Nearly half of the faint objects in the Antennae image are young clusters containing tens of thousands of stars. The orange blobs to the left and right of image centre are the two cores of the original galaxies and consist mainly of old stars criss-crossed by filaments of dust, which appear brown in the image. The two galaxies are dotted with blue star-forming regions surrounded by glowing hydrogen gas, appearing in the image in pink. During the course of the collision, billions of stars will be formed. The brightest and most compact of these star birth regions are called super star clusters. Collisions occur when two galaxies pass directly through each other and have sufficient relative momentum not to merge. The stars of interacting galaxies will usually not collide, but the gas and dust within the two forms will interact, sometimes triggering star formation

A starburst galaxy (Credit: ESA/(STScI/AURA)/Hubble collaboration/ NASA)

This image from Hubble Space Telescope in March 2006 shows throughout the galaxy's centre, young stars are being born 10 times faster than they are inside our entire Milky Way Galaxy. Most of the pale, white objects sprinkled around that look like fuzzy stars are actually individual star clusters about 20 light-years across and contain up to a million stars. Stars are created within galaxies from a reserve of cold gas that forms into giant molecular clouds. Some galaxies have been observed to form stars at an exceptional rate, which is known as a starburst. If they continue to do so, then they would consume their reserve of gas in a time span less than the lifespan of the galaxy

Active elliptical radio galaxy (Credit: Hubble Heritage Team/ NASA)

This image from Hubble Space Telescope on 6 July 2000 shows black hole-powered jet of electrons and sub-atomic particles streaming from the centre of the galaxy. A radio galaxy emits radio frequencies from relativistic jets. The mechanism for producing these jets is not well understood. A portion of the observable galaxies are classified as an active galaxy if the galaxy contains an active galactic nucleus. A significant portion of the total energy output from the galaxy is emitted by the active galactic nucleus, instead of the stars, dust and interstellar medium of the galaxy

Andromeda galaxy (Credit: NASA/JPL-Caltech)

This image from NASA's Galaxy Evolution Explorer on 15 May 2012 shows the ultraviolet side of Andromeda galaxy, our Milky Way's largest galactic neighbour. Andromeda is so bright and close to us that it is one of only ten galaxies that can be spotted from Earth with the naked eye. This view is two-colour composite, where blue represents far-ultraviolet light, and orange is near-ultraviolet light

Nebula is an interstellar cloud of dust, hydrogen, helium and other ionized gases. Originally, nebula was a name for any diffuse astronomical object, including galaxies beyond the Milky Way. Most nebulae are of vast size, even millions of light years in diameter and contrary to fictional depictions where starships hide in nebulae as thick as cloud banks, in reality a nebula that is barely visible to the human eye from Earth would appear larger, but no brighter, from close by. Although denser than the space surrounding them, most nebulae are far less dense than any vacuum created on Earth – a nebular cloud the size of the Earth would have a total mass of only a few kilograms. Many nebulae are visible due to their fluorescence caused by the embedded hot stars, while others are so diffuse they can only be detected with long exposures and special filters. Nebulae are often star-forming regions, such as in the "Pillars of Creation" in the Eagle Nebula (shown here from Hubble Space Telescope, Credit: NASA/Jeff Hester/ Paul Scowen (Arizona State University). In these regions the formations of gas, dust, and other materials clump together to form denser regions, which attract further matter, and eventually will become dense enough to form stars. The remaining material is then believed to form planets and other planetary system objects

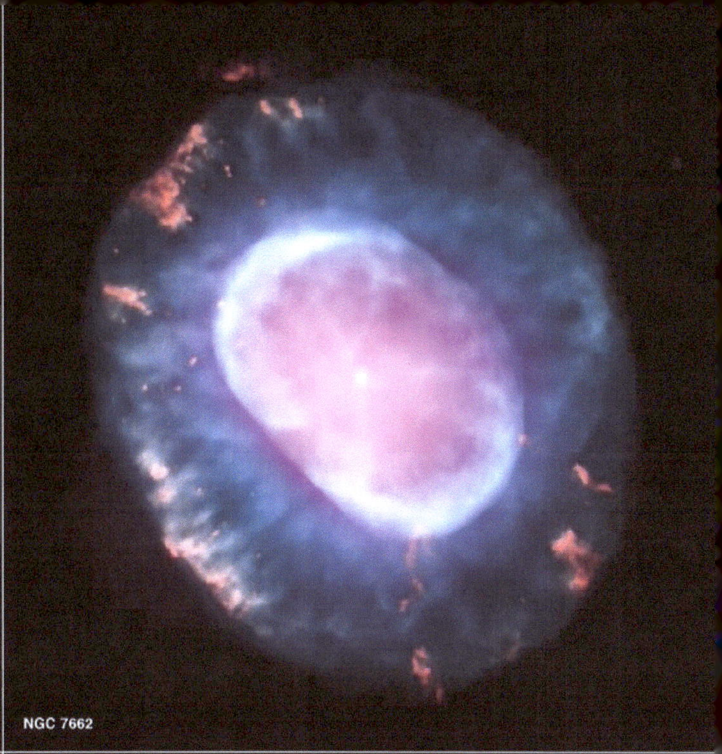

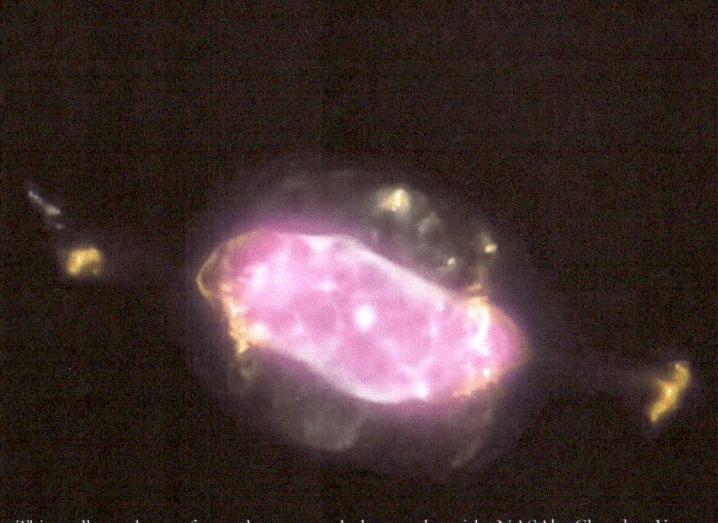

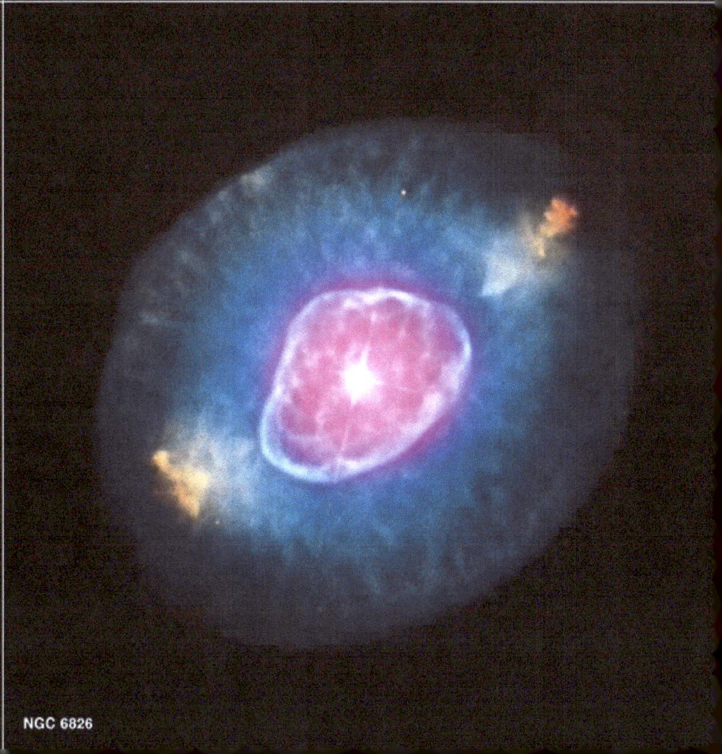

This gallery shows four planetary nebulas made with NASA's Chandra X-ray Observatory on 10 May, 2000 where X-ray emission from Chandra is coloured purple and optical emission from the Hubble Space Telescope is coloured red, green and blue. A planetary nebula results when a star like the Sun becomes a red giant and sheds its outer layers. Credit: X-ray: NASA/CXC/RIT/J.Kastner et al.; Optical: NASA/STScI

5. Universe
5.1 A brief background of Universe

The Universe is all of time and space and its contents including planets, moons, minor planets, stars, galaxies, the contents of intergalactic space, and all matter and energy and the physical laws that relate them

The Universe is everything that exists, everything that has existed, and everything that will exist encompassing all of life, all of history, and even ideas such as mathematics and logic

The spacetime of the Universe is usually interpreted with space as consisting of three dimensions, and time as consisting of one dimension (the fourth dimension)

The observable universe is about 28 billion parsecs (91 billion light-years) in diameter

The size of the entire Universe is unknown, but it is known that the Universe had a beginning and that it is expanding at an increasing rate

Because we cannot observe space beyond the edge of the observable universe, it is unknown whether the size of the Universe is finite or infinite. Even if we observe light traveled from the edge of the universe today, this light does not represent the distance at any given time because the edge of the observable universe and the Earth have since moved further apart. Due to this expansion, scientists on Earth can observe the light from a galaxy 30 billion light years away even though that light has traveled for only 13 billion years; the very space between them has expanded

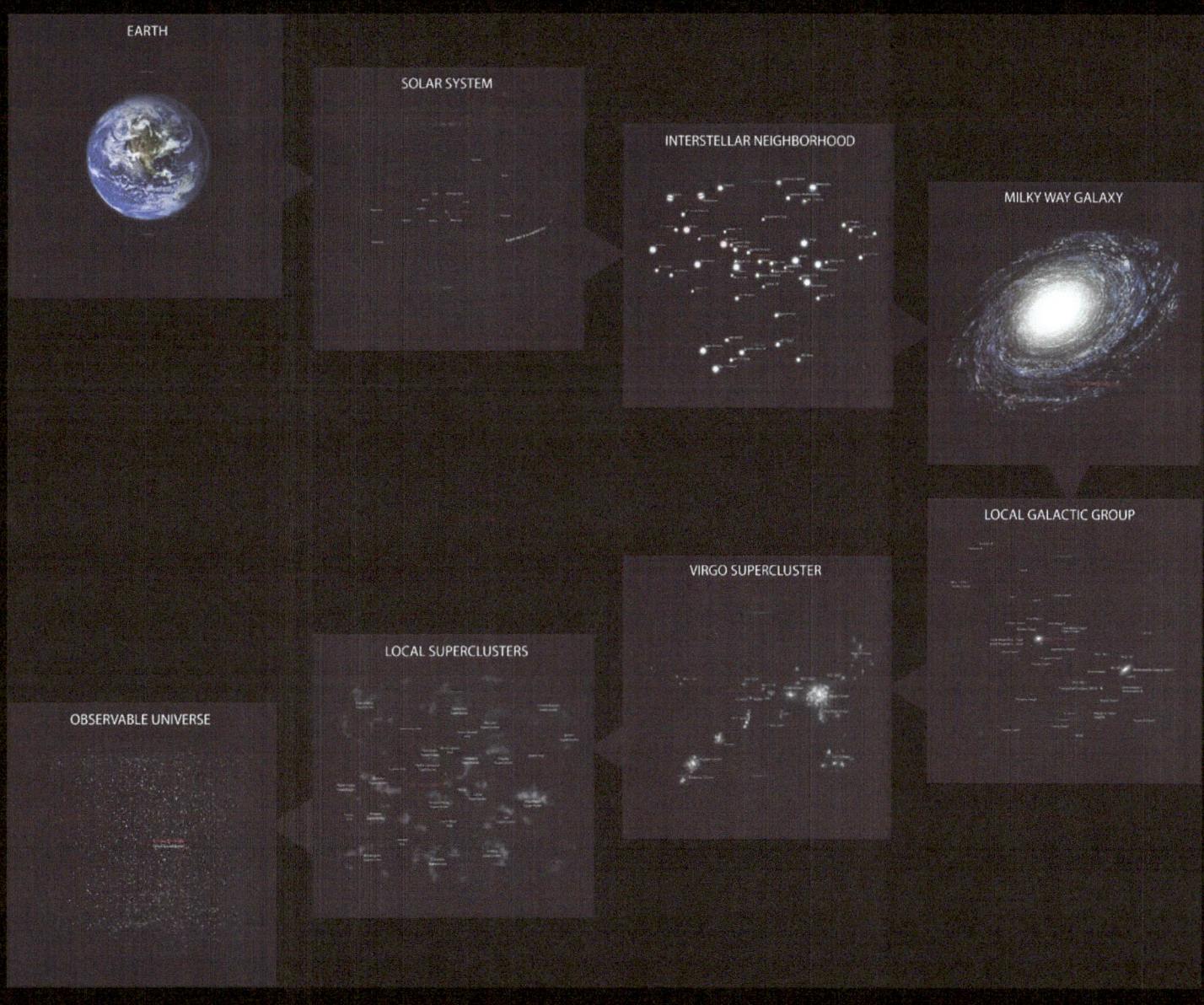

An artistic diagram of Earth's location in the Universe in a series of eight maps that show clockwise, starting with the Earth, moving to the Solar System, onto the Solar Interstellar Neighbourhood, onto the Milky Way, onto the Local Galactic Group, onto the Virgo Supercluster, onto our Local Superclusters, and finishing at the Observable Universe (Adapted from Andrew Z. Colvin)

5. Universe

5.2 The content of Universe

The Universe is composed almost completely of dark energy (68.3%), dark matter (26.8%), ordinary matter (4.9%), electromagnetic radiation (0.005%– 0.01%) and antimatter

Ordinary matter is, atoms, ions, electrons and the objects they form including stars, as well as interstellar gas in the interstellar and intergalactic media, planets, and all the objects from everyday life that we can bump into, touch or squeeze

Dark matter is a hypothetical kind of matter that is invisible to the entire electromagnetic spectrum, but which accounts for most of the matter in the Universe. Other than neutrinos, a form of hot dark matter, dark matter has not been detected directly and its existence and properties are inferred from its gravitational effects on visible matter, radiation, and the large-scale structure of the Universe. Dark matter neither emits nor absorbs light or any other electromagnetic radiation at any significant level

Dark energy is an unknown form of energy that is hypothesized to permeate space. It is the energy of empty space that is causing the expansion of the Universe to accelerate. The nature of both dark energy and dark matter is not well known

Electromagnetic radiation is the radiant energy released by certain electromagnetic processes. Visible light is an electromagnetic radiation. Its waves can be characterized by either the frequency or wavelength of the oscillations to form the electromagnetic spectrum, which includes, in order of increasing frequency and decreasing wavelength: radio waves, microwaves, infrared radiation, visible light (the only visible spectrum to human eyes), ultraviolet radiation, X-rays and gamma rays

Antimatter is a material composed of antiparticles, which have the same mass as particles of ordinary matter but opposite charges. Collision between particles and antiparticles leads to the annihilation of both and releases energy proportional to the total matter and antimatter mass. The Universe appears to have much more matter than antimatter. This asymmetry of matter and antimatter in the visible Universe is one of the great unsolved problems in physics

5. Universe
5.3 Studying the Universe

Astronomy, a natural science, is the study of celestial objects (such as stars, galaxies, planets, Moons, asteroids, comets and nebulae) and processes (such as supernovae explosions, gamma ray bursts, and cosmic microwave background radiation), the physics, chemistry, and evolution of such objects and processes, and more generally all phenomena that originate outside the atmosphere of Earth. Astronomy is the oldest of the natural sciences

Astronomy is one of the few sciences where amateurs can still play an active role, especially in the discovery and observation of transient phenomena. Amateur astronomers have made and contributed to many important astronomical discoveries, such as finding new comets

Cosmology is the study of the origin, evolution, and eventual fate of the universe. Physical cosmology is the scholarly and scientific study of the origin, evolution, large-scale structures and dynamics, and ultimate fate of the universe, as well as the scientific laws that govern these realities. The modern era of physical cosmology began in 1917, when Albert Einstein first applied his general theory of relativity to model the structure and dynamics of the Universe. Because of its shared scope with philosophy, theories in physical cosmology may include both scientific and non-scientific propositions, and may depend upon assumptions that cannot be tested. Modern physical cosmology is dominated by the Big Bang theory

The Big Bang theory, the prevailing cosmological model describing the development of the Universe, states that space and time were created in the Big Bang and were given a fixed amount of energy and matter that becomes less dense as space expands. According to the standard model of the Big Bang theory, the age of the Universe is measured to be 13.8 billion years. There are many competing hypotheses about the ultimate fate of the Universe and about what, if anything, preceded the Big Bang. The simplest Big Bang model accounts for observations such as the correlation of distance and redshift of galaxies (the ratio of the number of hydrogen to helium atoms), and the microwave radiation background

Mankind's deepest-ever view of the Universe (Credit: NASA/ESA/G. Illingworth/D. Magee/P. Oesch (University of California, Santa Cruz)/R. Bouwens (Leiden University)/HUDF09 Team)

This image called eXtreme Deep Field (XDF), was assembled by combining 10 years of Hubble Space Telescope images. By collecting faint light over many hours of observation, it revealed thousands of galaxies, both nearby and very distant, making it the deepest image of the universe ever taken at that time. The XDF contains about 5,500 galaxies even within its smaller field of view. The faintest galaxies are one ten-billionth the brightness of what the human eye can see

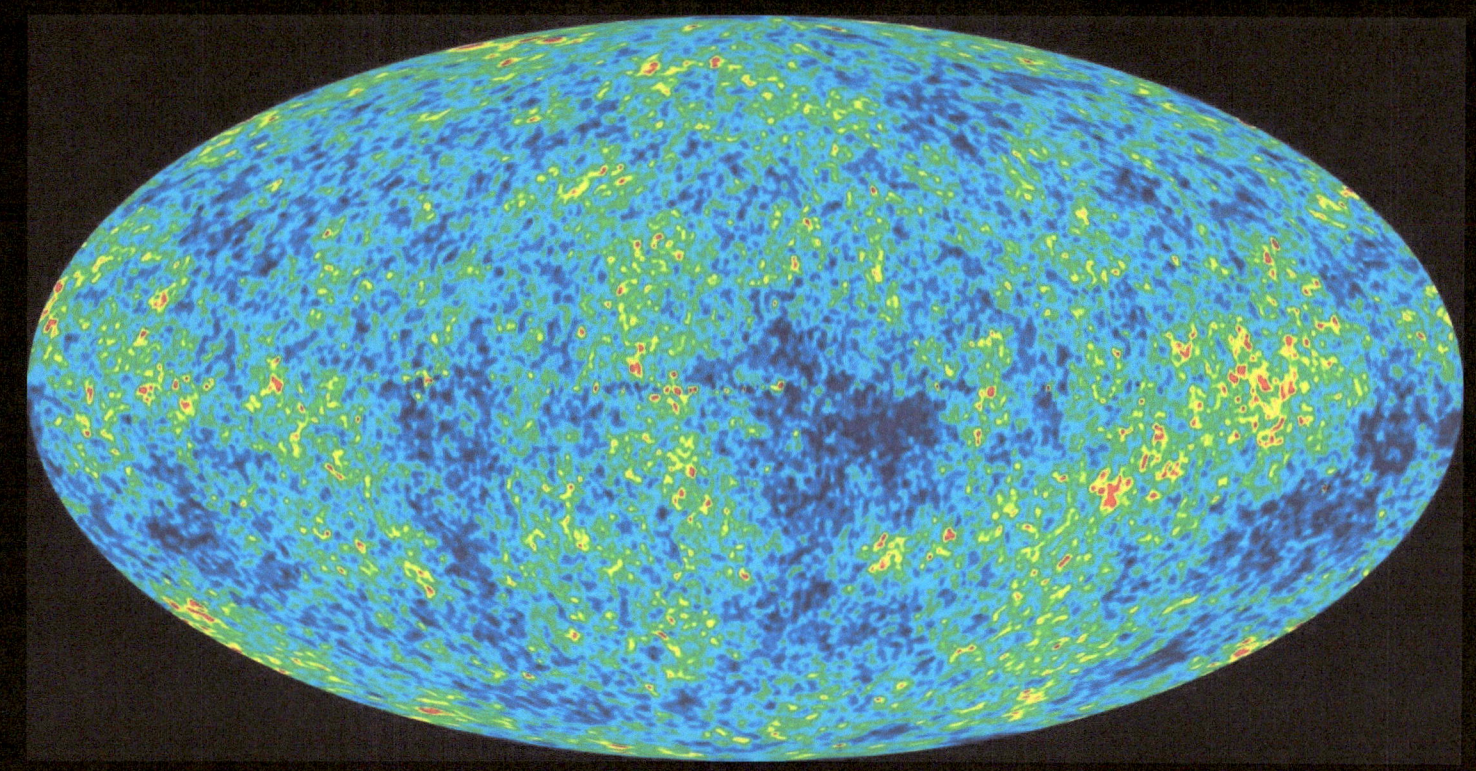

Nine Year Microwave Sky from WMAP (Credit: NASA / WMAP Science Team)

The detailed, all-sky image of the infant universe created from nine years of WMAP data. The image reveals 13.77 billion year old temperature fluctuations (shown as colour differences) that correspond to the seeds that grew to become the galaxies. The signal from our Galaxy was subtracted using the multi-frequency data. This image shows a temperature range of ± 200 microKelvin. The Wilkinson Microwave Anisotropy Probe (WMAP) was a spacecraft operating from 2001 to 2010 which measured temperature differences across the sky in the cosmic microwave background (CMB) – the radiant heat remaining from the Big Bang. CMB is a cosmic background radiation that is fundamental to observational cosmology because it is the oldest light in the universe, dating back to when the Universe was just 380,000 years old. With a traditional optical telescope, the space between stars and galaxies (the background) is completely dark. However, a sufficiently sensitive radio telescope shows a faint background glow, almost isotropic, that is not associated with any star, galaxy, or other object. This glow is strongest in the microwave region of the radio spectrum

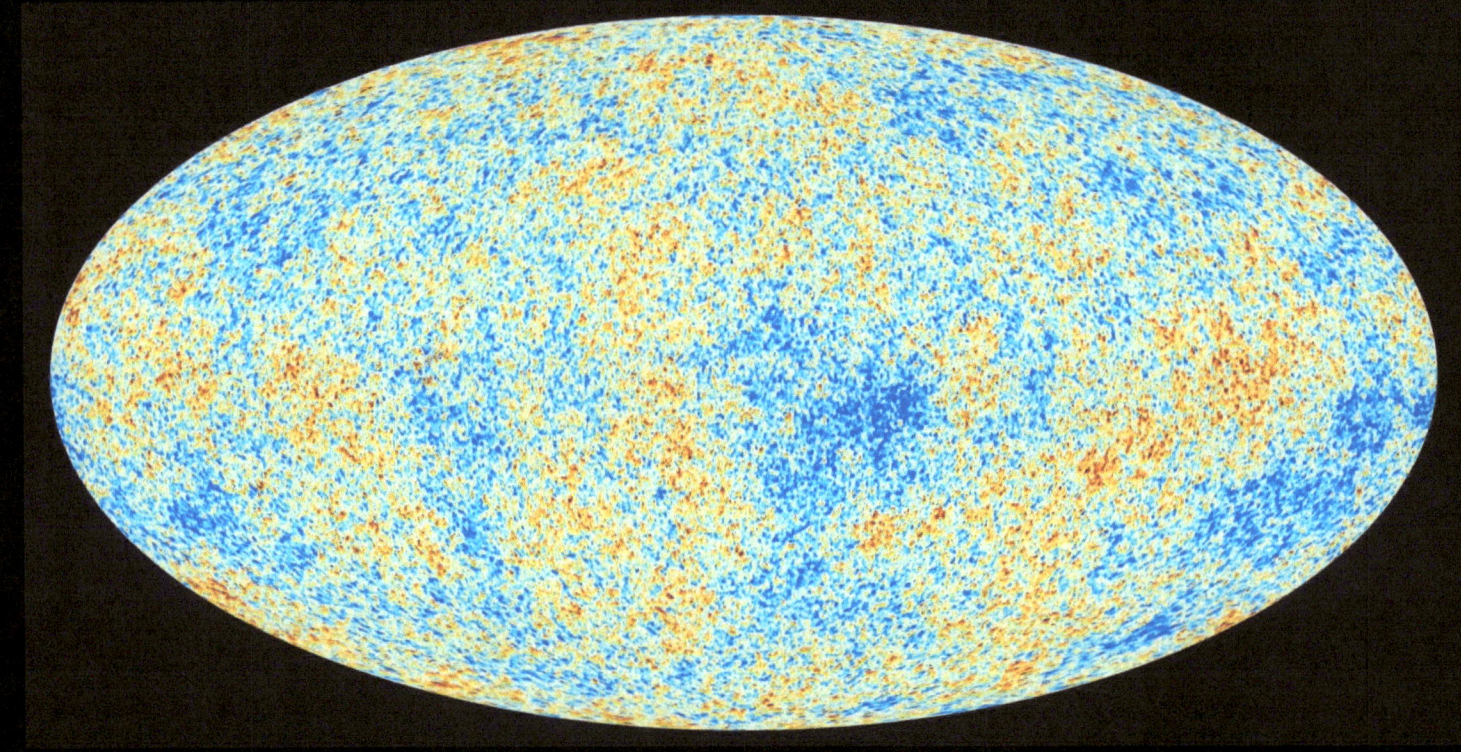

Microwave Sky from Planck cosmology space probe (Credit: ESA/Planck Collaboration)

An all-sky map of the cosmic microwave background (CMB) from Planck cosmology space probe. This map suggests the Universe is slightly older than thought: according to the map, subtle fluctuations in temperature were imprinted on the deep sky when the Universe was about 370,000 years old. The map shows tiny temperature fluctuations that correspond to regions of slightly different densities, representing the seeds of all future structures: the stars and galaxies of today. The imprint also reflects ripples that arose as early in the existence of the Universe as the first nonillionth (10^{-30}) of a second. It is currently theorised that these ripples gave rise to the present vast cosmic web of galactic clusters and dark matter. According to the Planck Collaboration team, the Universe is 13.798 billion years old, and contains 4.82% ordinary matter, 25.8% dark matter and 69% dark energy. Red areas are slightly warmer than average, and blue areas are slightly cooler

Microwave Sky from Planck cosmology space probe (Credit: ESA/LFI & HFI Consortia)

This multi-frequency all-sky image of the microwave sky has been composed using data from Planck cosmology space probe covering the electromagnetic spectrum from 30 GHz to 857 GHz. The mottled structure of the cosmic microwave background (CMB), with its tiny temperature fluctuations reflecting the primordial density variations from which today's cosmic structure originated, is clearly visible in the high-latitude regions of the map. The central band is the plane of our Galaxy. A large portion of the image is dominated by the diffuse emission from its gas and dust. To the right of the image, below the plane of the Galaxy, is a large cloud of gas in our Galaxy. The obvious arc of light surrounding it is Barnard's Loop – the expanding bubble of an exploded star. Planck has seen whole other galaxies. The great spiral galaxy in Andromeda, 2.2 million light-years from Earth, appears as a sliver of microwave light, released by the coldest dust in its giant body. Other, more distant, galaxies with super massive black holes appear as single points of microwaves dotting the image

A Milky Way 'Mixer' Amongst the Stars (Credit: ESA/NASA/JPL-Caltech)

Usually, images of the microwave sky are composed using data from cosmology space probes covering the electromagnetic spectrum from 30 GHz to 857 GHz by subtracting the Milky Way's light to isolate the ancient signals. However, here the Planck cosmology space probe image is constructed from observations made at microwave and millimetre wavelengths to portray our Milky Way galaxy. Red shows dust that gives off a thermal glow. Yellow shows carbon monoxide gas, which is concentrated along the plane of our Milky Way in the densest clouds of gas and dust that are churning out new stars. Blue indicates a type of radiation called synchrotron, which occurs when fast-moving electrons, spit out of supernovas and other energetic phenomena, are captured in the galaxy's magnetic field. The green shows a different kind of radiation known as free/free which occurs when isolated electrons and protons careen past one another in a series of near collisions, slowing down but continuing on their own way (the name free/free comes from the fact that the particles start out alone and end up alone). The free/free signatures are associated with hot, ionized gas near massive stars

Game of Stars

1. Explain this saying to your friends: "when we look out in space, we look back in time"
2. What is the difference between false and true colour images?
3. Scientists are so reluctant to place ordinary high definition cameras on spacecraft or probe to take natural colour images. Why?
4. Discuss at least one primary cause of Earth's ice ages
5. How can you explain the mass extinctions Earth has experienced in 100s of millions of years intervals as a possible result of Solar System's position in Milky Way?
6. There is a growing trend in people worrying about the health effects of radio waves used for cellular phone communications. Discuss for and against the health effect of cell phone radio waves
7. Earth is unlikely to be hit by a large celestial body originating from outside our Solar System. Why?
8. Which one of the three Earth-like planets experienced extreme climate change impact?
9. When I was a kid, Pluto was a ninth planet of our planetary system. Why is Pluto not a planet anymore?
10. Which of the two planets of the Solar System lose more energy than they get from the Sun?
11. The Earth is literally falling towards the Sun under its immense gravity but why doesn't the Earth fall down onto the Sun's surface? While you are there, why the spacecrafts orbiting Earth do not fall down onto the Earth surface?
12. What are the coldest and hottest planets of our Solar System?
13. Jupiter is so big that theoretically it is the maximum size a planet can get before it becomes a star. Why do you think Jupiter may never become a star?
14. Which of our planets or planetary system objects resemble Sun in size and composition?
15. What is the colour of the Sun?
16. How can you tell the age of a star?
17. Why are all stars fixed in space?
18. Among the contents which form the Universe which is the most abundant?
19. What are the four dimensions of space-time of the Universe?
20. The age of the Universe is 13.8 billion years but scientists can observe the light from a galaxy 30 billion light years away even though that light has travelled for only 13 billion years. How is this possible?
21. There are so many unknowns about our Universe including the nature of dark energy, dark matter, antimatter, black holes to mention but a few. What is the primary way of knowing the existence and nature of celestial bodies which never emit or reflect light, or do not let anything escape from inside them including particles and electromagnetic radiation such as light?

ABOUT THE AUTHOR

Alemu Gonsamo (PhD in Earth Observation, 2010), is a research scientist at the University of Toronto. He has published more than 60 peer reviewed research articles in the broad field of Earth observation. He serves as a reviewer for 30 peer reviewed international journals in various areas of Earth System Sciences. Alemu has been invited speaker at Cornell University, Chinese Academy of Sciences, University of California - Davis, among others, and serves as a grant review board member for European Commission and NASA. He lives in Toronto, Ontario, Canada